The Electrification of Russia,
1880–1926

The
Electrification
of Russia,
1880–1926

Jonathan Coopersmith

Cornell University Press

ITHACA AND LONDON

First published 1992 by Cornell University Press.

Library of Congress Cataloging-in-Publication Data

Coopersmith, Jonathan, 1955–
    The electrification of Russia, 1880–1926 / Jonathan Coopersmith.
        p.   cm.
    Includes bibliographical references (p.   ) and index.
    ISBN 0-8014-2723-1 (alk. paper)
    1. Electrification—Economic aspects—Soviet Union. 2. Electric industries—Soviet Union—History. 3. Soviet Union—Economic conditions—1861–1917. 4. Soviet Union—Economic conditions—1918–1945. I. Title.
    HD9685.S762C66 1992
    333.79'32'0947—dc20   92-5735

Printed in the United States of America

In memory of my father

Contents

Illustrations

Acknowledgments

Like the story it tells, this book has a long history. My research began at Oxford University under the direction of Michael Kaser and Margaret Gowing; I also benefited from a year at the University of Pennsylvania with Thomas P. Hughes. Since then, an International Research and Exchange Board fellowship gave me an invaluable year in Soviet archives and libraries. Further support came from an Institute of Electrical and Electronic Engineers Postgraduate Fellowship, which enabled me to spend a year at the MIT Program in Science, Technology, and Society and the Harvard Russian Research Center, and a Title VIII fellowship of the Department of State, which provided a year at the Hoover Institution (Discretionary Grant Program, Department of State, Soviet-Eastern European Research & Training Act of 1983, Public Law 98–164, Title VIII, 97 Stat. 1047–50). American taxpayers and electrical engineers have contributed more to this book than they know. I hope they find their support justified.

This book has benefited greatly from the kindness of colleagues who read draft sections. For their patience and thoughts, I thank Harley D. Balzer, W. Bernard Carlson, Alek G. Cummins, Elihu Gerson, Lisa A. Halperin, Michael F. Hamm, David Hochman, Alfred Imhoff, Paul Josephson, Walther Kirchner, Thomas J. Misa, Thomas C. Owen, Orest Pelech, and Edmund N. Todd.

My gratitude also goes to Soviet archivists and the librarians of BAN (Library of the Academy of Sciences), the British Museum, the Centre for Russian and East European Studies, University of Birmingham, Harvard University, the Hoover Institution, the Institut Nauchnoi Informatsii po Obshchestvennum Naukam, the Lenin Library, the Library of Congress, MIT, Stanford University, the University of Illinois, Urbana-Champaign, and the Texas A&M Interlibrary Loan Office. Colleagues and staff at the Texas A&M History Department provided a supportive environment. In addition to the graphics, Lisa Halperin provided much encouragement through the many moves, missed deadlines, and numerous drafts.

The historians and librarians who helped me gave proof to the meaning of the academic community. One of the pleasures in writing this book has been the overlap between the academic world and my community of friends. Although ultimately an individual work, this book progressed because of the support of family, friends, and colleagues.

Portions of the book draw on two of my articles, and I am grateful to the publishers for permission to make use of this material. "The Role of the Military in the Electrification of Russia, 1870–1890" was published in Everett Mendelsohn, M. R. Smith, and Peter Weingart, eds., *Science, Technology, and the Military* 13 (1988): 291–305; copyright © 1988 by Kluwer Academic Publishers and used by permission of Kluwer Academic Publishers. "Technology Transfer in Russian Electrification, 1870–1925" appeared in *History of Technology* 13 (1991): 214–33 and is used by permission of Mansell Publishing.

JONATHAN COOPERSMITH

College Station, Texas

Abbreviations and
Russian Terms

I HAVE TRANSLITERATED RUSSIAN WORDS according to the Library of Congress system with several minor exceptions: I omit soft and hard signs and use the spelling of Russian names known in Western Europe instead of a modern transliteration (e.g., Jablochkov, not Iablochkov). I have modernized prerevolutionary spelling.

AC: alternating current
Artkom: artillery committee of the GAU
Azneft: Azerbaijan Unified Oil Industry
DC: direct current
duma: legislative council of a city
Elektrostroi: Utility construction section of the KGS; contained the TsES
Elektrotok: Leningrad Unified State Electric Stations, formerly Petrotok; also the ETO subsection for electric stations
ETO: Electrotechnical Section of the VSNKh
GAU: Main Artillery Administration (Glavnoe artilleriiskoe upravlenie)
Glavelektro: Main Administration for Electrotechnology; replaced the ETO
GOELRO: State Commission for the Electrification of Russia (Gosudarstvennaia komissiia po elektrifikatsiiu Rossii)
Gosplan: State Planning Commission (Gosudarstvennyi plan)
gradonalchik: appointed mayor
gubernator: appointed administrator of a region
guberniia: province
GUKKh: Main Administration for Cities (Glavnoe upravlenie kommunalnogo khoziaistva) of the NKVD
IRTO: Imperial Russian Technical Society (Imperatorskoe russkoe tekhnicheskoe obshchestvo)
ispolkom: executive committee (*ispolnitelnyi komitet*)

xi

KGS: Committee for State Construction (Komitet gosudarstvennykh sooruzhenii)

khozraschet: financial self-sufficiency

kustar industry: handicraft industry

kV: kilovolt

kW: kilowatt

kWh: kilowatt-hour

mazut: heating oil

MkWh: million kilowatt-hours

MOGES: Moscow Unified State Electric Stations (Moskovskoe obedinenie gosudarstvennykh elektricheskikh stantsii)

MTP: Ministry of Trade and Industry (Ministerstvo torgovli i promyshlennosti), before 1905 a department in the Ministry of Finance

MVD: Ministry of Internal Affairs (Ministerstvo vnutrennikh del)

MW: megawatt

NEP: New Economic Policy (Novaia ekonomicheskaia politika)

NKVD: People's Commissariat of Internal Affairs (Narodnyi komissariat vnutrennikh del)

Petrotok: predecessor of Elektrotok

remeslo industry: handicraft industry

SED: Council for Electrotechnical Affairs (Soveshchenie po elektrotekhnicheskim delam) of the MTP

SNK: Council of People's Commissariats (Sovet narodnykh komissarov)

TsES: Central Electrotechnical Council (Tsentralnyi elektrotekhnicheskii sovet) under the KGS and, later, the Main Administration for State Construction

TsVPK: Central War Industries Committee (Tsentralnyi voenno-promyshlennyi komitet)

uprava: executive administrative board of a city

VSNKh: Supreme Council for the National Economy (Vysshii sovet narodnogo khoziaistva)

zemstvo: local rural government

Footnote abbreviations

SRE: Sdelaem Rossiiu elektricheskoi. Moscow: Gosenergizdat, 1961

TsGANKh: Central State Historical Archive of the National Economy

TsGIAL: Central State Historical Archive of Leningrad

TsGIAMO: Central State Historical Archive of the Moscow Region

TsGVIA: Central State Military Historical Archive

ZIRTO: Zapiski Imperatorskogo russkogo tekhnicheskogo obshchestva

The Electrification of Russia,
1880–1926

CHAPTER 1

Introduction: The Shaping of a Technology

In December 1920, electrical engineer and Bolshevik Gleb M. Krzhizhanovskii displayed an illuminated map of a future electrified Russia to convince the 8th Congress of Soviets to approve a plan for state electrification. Moscow's generating capacity was so low, however, that lighting the bulbs on the map resulted in blacking out parts of the city. Electrification had great political significance for the Communist regime, but dreams outpaced reality.

As well as changing night into day, electrification transformed capital markets, the military, manufacturing, the spatial geography of cities, and many other facets of Russian life. One of the products of the industrial revolution beginning in the last third of the nineteenth century, electrification was a science-based high technology that demanded educated technicians and scientists as well as enormous amounts of capital and industrial capability. In 1920, electrification replaced the railroad as the state technology by which the new government intended to accomplish its political and economic goals and distinguish itself from the old government. How effective was the early Soviet Union in implementing this vision? The answer lies between 1880 and 1926, between the formation of Russia's first electrotechnical society and the initial steps toward Stalin's superindustrialization.

The approach used in this book is based on a social construction of technology, a powerful analytic tool that deepens our understandings of technologies and the societies in which they are grounded. Technology is viewed not as a given but as part of a "seamless web" with society.[1] To distinguish among science, technology, economics, and society is to create false dichotomies. Indeed, successful entrepreneurs are coalition builders who can forge alliances between their technologies and important social, economic, and political groups to gain support and resources.[2] And important as individuals and technologies are, their interactions are mediated by organizations, themselves constructs. As Thomas P. Hughes, Michel Callon, and other historians and sociologists of technology have demonstrated, "organizations as well as physical artifacts have to be invented."[3] Technological controversy, it should be noted, is normal and serves as an arena for competing economic, political, and social interests.[4]

In this history I investigate how political, economic, and social factors shaped the development of electrification in Russia and how electrification affected Russian society. I use Clifford Geertz's "thick" description and the broad characteristics of the political and economic strands inherent in tsarist and Soviet infrastructures.[5] The viewpoint is that of the engineers and technologists who struggled to form alliances to promote particular visions of electrification. Economic development and political factors structure technological change by guiding institutions and individuals along certain paths and excluding others. As I show, these institutional arrangements did not occur by chance but resulted from political, bureaucratic, and economic struggles among competing groups. These struggles and the concomitant battles for resources and prestige shaped the development of Russian electrification more than did technological momentum.

The historical interest in the rapid economic and industrial development of prerevolutionary Russia during its last decades has concentrated on the technologies of the first industrial revolution: iron and steel, textiles, and railroads. Industrialization meant railroads, not power plants, and tsarist resources (and our historical attention) were

[1] "Introduction," in Wiebe E. Bijker, Thomas P. Hughes, and Trevor J. Pinch, eds., *The Social Construction of Technological Systems* (Cambridge: MIT Press, 1987), 10.

[2] See Harvey M. Sapolsky, *The Polaris System Development: Bureaucratic and Programmatic Success in Government* (Cambridge: Harvard University Press, 1972), and Bruno Latour, *Science in Action: How to Follow Scientists and Engineers through Society* (Cambridge: Harvard University Press, 1987).

[3] "Introduction," and Michel Callon, "Society in the Making: The Study of Technology as a Tool for Sociological Analysis," in Bijker, Hughes, and Pinch, eds., *Social Construction*, 14 and 83–102.

[4] See Dorothy Nelkin, ed., *Controversy: Politics of Technical Decisions*, 2d ed. (Beverly Hills, Calif.: Sage, 1984).

[5] Clifford Geertz, "Thick Description: Toward an Interpretive Theory of Culture," in *The Interpretation of Cultures* (New York: Basic Books, 1973), 3–30.

focused on the iron horse. American research on Russian technology has been devoted to the Soviet era and rarely touches on electrotechnology. The voluminous Soviet literature on electrification concentrates on the post-1920 period, when electrification became a state technology under Lenin's slogan "Communism is Soviet power plus the electrification of the whole country."[6]

How should we understand electrification in the nation-state context? Comparing the European and American experiences, Thomas P. Hughes suggests that electrification follows a pattern from invention to development, innovation, transfer, growth, competition, consolidation, and rationalization.[7] Each stage involves different people and institutions. Throughout, the successful entrepreneur adapts the technology to its environment while harnessing outside political and economic forces. Hughes concentrates on Western industrialized states, which are not representative of most of the world. From an industrializing viewpoint, Hughes overemphasizes invention, underemphasizes diffusion, and neglects questions of technology transfer and infrastructure. As post-Meiji Japan has demonstrated, invention is important but not vital for industrialization. Although invention has justly received a great deal of attention from historians, transfer is the key step in economic development. In the case of Russia, problems in transferring electrotechnical technology—institutional, economic, and other—slowed the spread and growth of Russian utilities and, consequently, of modernization.

How did electrification in Russia differ from that in the West?[8] Why did the pace of electrification in Russia proceed so slowly or, an equally appropriate question, how did the West electrify so quickly? Russian electrical engineers did not work in isolation but actively participated in the international electrotechnical community. Commercial development of electricity for light, power, and transportation proceeded more rapidly in the West, however, and Russia never caught up. Major Russian differences included a lower level of urbanization and industrialization, a lack of supportive financial and technical infrastructure, and an overarching state administrative framework.

Electrification was one of four network technologies to transform

---

[6] *Vosmoi vserossiiskii sezd sovetov: Stenograficheskii otchet* (Moscow: Gosudarstvennoe Izdatelstvo, 1921), 30.

[7] Thomas P. Hughes, "The Evolution of Large Technological Systems," in Bijker, Hughes, and Pinch, eds., *Social Construction*, 57–73.

[8] The "West" is an ambiguous term, often used as a higher standard of comparison with Russia. Here, it refers to the more technologically advanced countries Germany, Great Britain, France, and the United States.

the urban environment in the late nineteenth century; the others were communications (telephone and telegraph), transportation (trains, trams, buses, and automobiles), and health (water and sewage). In a network technology, individual components—for electricity, generating stations, transmission wires, control and distribution systems— do not work unless the whole system functions. These technologies are usually considered natural monopolies because of the large amounts of capital required and the accompanying centralized control. Instead of resulting from the operation of two parallel systems, competition primarily occurs in the proposal stage when entry costs are low. The large expense of these technologies, their major role in a modern economy, and the political negotiations and alliances necessary to build, expand, and operate their networks means, as Josef W. Konvitz put it, that "the nature and extent of the controls built into systems and imposed upon them from outside mattered as much [as] or more than any purely technical factors."[9]

The history of electrification can be viewed as one of "bigger and better": more area covered by a single grid; larger power stations to service larger areas; more intensive use of electricity for light, power, and traction; and increased control exercised over the entire system in the name of more rational and efficient operations. Another interpretation is possible, however, one that sees the so-called natural aspect of monopoly determined as much by social and economic factors as by technical considerations. What is natural for a particular technology in a particular country at a particular time depends on the environment. Electrification became a natural activity in Russia in a different way from its development in the West.

Electrification was revolutionary worldwide, but it was also conventional. Electricity does little that cannot be done by other technologies and energy sources. Kerosene and gas provide lighting; wood and coal supply heat; belt drives transmit energy; horses pull trams. In its simplest applications, electricity replaced these sources; in its more advanced and developed forms, conceived within a broader systemic context, electricity could radically transform a workplace, an industry, and even a nation. The economic and social importance of electrification grew as its uses expanded from a novel means of lighting in the 1880s to industrial applications and trams in the 1890s. By World War

[9] Josef W. Konvitz, *The Urban Millennium: The City-Building Process from the Early Middle Ages to the Present* (Carbondale: Southern Illinois University Press, 1985), 135. See also Joel A. Tarr and Gabriel Dupuy, eds., *Technology and the Rise of the Networked City in Europe and America* (Philadelphia: Temple University Press, 1988).

I, urban society and heavy industry depended on electric energy. The original electric companies combined the functions of generation, transmission, and distribution in small areas. As utilities grew, they remained in control of all three functions, seeing them as one continuous operation. By World War I, the scale and importance of electric light, power, and traction were such that alternative approaches for utility organization and control increased in number and credibility in Russia, Europe, and North America. After the war, proposals for centralized, large-scale, regional electrification received growing attention as part of a technocratic movement by engineers. Only in Russia, however, did the economic and political upheavals that started in 1914 change the status of electrification as well as the government.

Three themes flow through this book—the omnipresent foreign role in Russian electrification, the political constituency for the electrification process, and the economic, technical, and administrative environment in which it was attempted. Understanding Russian electrification is impossible without including the large foreign influence to which it was subject, influences that ranged from the obvious financial and material transfers to the less overt but very important flows of ideas, legitimization, and people. The lack of a national political constituency resulting from the centralized nature of the state handicapped electrification in tsarist Russia, but similar Soviet centralization benefited the politically connected electrical engineers. In general, the economic, technical, and administrative environment encompasses an invisible supporting infrastructure of activities from capital availability to government regulations to trained personnel. As electrification's slow prerevolutionary transfer and diffusion illustrate, environment plays a major role in the development of a technology, one noticed more when it hinders than when it helps.

To explore these issues, I divide this book into three chronological parts: 1880–1914, the last years of imperial Russia; 1914–20, the chaotic years of world and civil war; and 1920–26, the period of the New Economic Policy (NEP). In each part I discuss the political and economic environment, the main actors, and the legal and organizational foundations of electrification. I exclude electrical manufacturing and the electrification of industry, which have been explored elsewhere.[10]

---

[10] V. S. Diakin, *Germanskie Kapitali v Rossii* (Leningrad: Nauka, 1971); S. A. Gusev, *Razvitie sovetskoi elektrotekhnicheskoi promyshlennosti* (Moscow: Energiia, 1964); Guenter S. Holzer, "German Electrical Industry in Russia" (Ph.D. diss., University of Nebraska, 1970); Walther Kirchner, "The Industrialization of Russia and the Siemens Firms 1853–1890," *Jahrbucher für Geschichte Osteuropas* 22 (1974): 321–57, "Siemens and AEG and the

The focus is on the development of central and regional stations, which provided electricity for residential, commercial, government, and industrial users.

Chapters 2 and 3 on imperial Russia cover the initial decades of electric light, power, and traction as utilities spread from St. Petersburg and Moscow to other cities and towns. Although generation and transmission technologies evolved rapidly, development followed a conservative pattern, as suggested by the predominance of direct over alternating current.

Chapters 4 and 5 on World War I, the 1917 revolutions, and the civil war cover a time of profound change for utilities and electrical engineers. During this period, electrification changed from a local concern to a matter of vital importance and promise to the state. As in the West, World War I served as a catalyst for increased economic centralization and control in Russia, but six years of war and revolution created a political situation in which electrification became the new state technology par excellence. Aided by an increasingly dire economic and revolutionary environment, radical plans for utility development gained support among engineering and political elites. The core of this radical thinking was the regional station, a single powerplant that could serve hundreds of square kilometers.

Chapters 6 and 7 cover the State Commission for the Electrification of Russia (GOELRO) plan for state electrification and its partial implementation during the NEP. The harsh realities of reconstruction and limited resources overwhelmed the optimistic hopes of planners, who had to fend off demands from urban utilities and more radical proposals for rapid rural electrification.

As electric lighting, power, and traction grew in economic importance in Russia, their political importance also increased. Before 1914, electrification received no special treatment from the state; after 1920, it was the state technology, supported by the government and Communist party as a means to achieve their ends. As with so many other aspects of society, the tsarist administration's "normal" treatment slowed the development and diffusion of new technologies and businesses. In the postrevolutionary period, the prominence given to electrification shows the meshing of agendas of different groups—the leadership of the electrical engineers and the Communist party—produced a mutually beneficial program at the expense of alternatives.

Electrification of Russia, 1890–1914," *Jahrbucher für Geschichte Osteuropas* 30 (1982): 399–428, and *Die Deutsche Industrie und die Industrialisierung Russlands, 1815–1914* (St. Katharinen: Scripta Mercaturae Verlag, 1986).

Postrevolutionary planners had three technological choices, each with a different set of political, economic, and social assumptions and priorities. The possible paths were a conservative approach, desired by cities, of supporting their existing utilities; a radical approach of rapid rural electrification supported by political and engineering advocates of social transformation; and a centralized approach of regional stations first for Moscow and Leningrad and later for other industrial centers promoted by engineers, planners, and Communists with a technocratic bent. The Communist party chose the third and technically most demanding approach despite opposition from advocates of radical and rapid decentralized rural electrification and proponents of existing medium-scale urban utilities. Although justified on technocratic criteria of maximizing economic rationalization and industrial development, the decision was inherently political. The importance of electrification ensured that authority over its development rested not in the leadership of the electrical engineering community but in the Communist party. Although electrical engineers occupied important government positions, they discovered that their monopoly on technical expertise did not give them a monopoly on decision making and resource allocation.

This history ends in 1926, when electrification had recovered from the travails of 1914–20 but before it was subordinated to industrialization. The political shift in 1925–26 from support of electrification to support of industrialization, the rapid growth of regional stations after 1927, and the advent of the five-year plans produced a style of electrification quite different from that of the early NEP years. After 1926, we must talk not about electrification in Russia but about Soviet electrification, which has a history sufficiently complex to warrant its own volume.

CHAPTER 2

Government and Growth in
Imperial Russia, 1870–1886

THE DEVELOPMENT OF ELECTRIFICATION exemplifies the transfer and
diffusion of a new technology into Russian society and the growing
technological gap between Russia and the West. Electric lighting,
power, and traction advanced greatly, but their geographic diffusion
and intensity of application trailed the West's. In this chapter I explore five key factors that shaped prerevolutionary electrification: the
restrictive institutional environment imposed by the tsarist government, the strong military role, the weak commercial reception of native invention, the development of the electrical engineering community, and significant foreign financial and technical involvement.

The administrative and legal environment of tsarist electrification
helps explain why the Russian economy proved less supportive of
electrotechnology than did Western European and American economies. Electrification suffered, as did most economic activities in
Russia, from the government's restrictive procedures. The Russian
army and navy were significant exceptions to this government indifference, and thus they played the major role in the initial establishment of electrotechnology. The general failure of inventors in Russia
illuminates the weak social and institutional support for technological
innovation, innovation supplied later by foreign firms. In both military and civilian spheres, electrical engineering societies tied Russian
electrotechnology together. Never passive, electrical engineers gradually became involved in the politics of electrification. Their full involvement, however, came about only when World War I radically
changed the political and economic environment.

8

The Role of Government

GOVERNMENTS SHAPE THE DEVELOPMENT, diffusion, and evolution of new technologies by, among other factors, their approach to risk, access to funding, decision making, and markets.[1] Budget priorities, tax structures, regulation, political favoritism, national security, and other goals of elites in power can aid, deliberately or otherwise, certain technologies while hindering others.

In circumstances of "business as usual," a new technology evolves within an established framework of precedent, regulation, and authority. The government neither gives the new technology special benefits nor penalizes it. Sometimes a government actively promotes one technology at the expense of other options for military, economic, political, and social goals. Such a state technology is supported directly and publicly as the government identifies itself with that technology. Similarly, supporters of that technology try to place themselves under government aegis. These technologies tend to be capital-intensive, regional in scope, and monopolistic, and they generally strengthen the central powers of the state. Although there is no intrinsic reason why state technologies must be capital-intensive high technologies (e.g., nuclear reactors instead of solar water heaters), the demands for large amounts of technical and economic resources, coupled with the centralizing tendencies of both the state and that technological approach, provide an alluring combination. The railroad and the space program are two examples. Railroads were revolutionary forces of modernization that helped solidify the nation-state as a political and economic entity. Governments promoted railroads to develop national markets, steel and manufacturing industries, and financial institutions while strengthening their military power.[2] Governments supported and guided domestic and international space programs, like railroads, for reasons of national security, political prestige, and economic and technological development.[3]

---

[1] Nathan Rosenberg and L. E. Birdzell, *How the West Grew Rich: The Economic Transformation of the Industrial World* (New York: Basic Books, 1986), 24–32; Thomas C. Cochran, *Frontiers of Change: Early Industrialization in America* (New York: Oxford University Press, 1981), 39, 121.

[2] Alfred D. Chandler, Jr., *The Visible Hand* (Cambridge: Harvard University Press, 1977), 79–187; Dennis E. Showalter, *Railroads and Rifles: Soldiers, Technology and the Unification of Germany* (Hamden, Conn.: Archon, 1976); Eugen Weber, *Frenchmen into Peasants* (Stanford: Stanford University Press, 1976), 205–6.

[3] John M. Logsdon, *The Decision to Go to the Moon: Project Apollo and the National Interest* (Chicago: University of Chicago Press, 1970); Bruce Mazlish, ed., *The Railroad and the Space Program: An Exploration in Historical Analogy* (Cambridge: MIT Press, 1964); Walter

In developing a state technology, a government seeks to strengthen its economy while simultaneously increasing its domestic and international political standing.[4] In an economist's ideal market, governments distort the natural development of a new technology by promoting one technology over others. In reality, state actions are part of the normal development of a technology. As technology is identified with progress, economic growth, and military superiority, governments link themselves with it.

State technologies are marriages of convenience in which the promoters of a technology join with the government to pursue common interests, albeit for different reasons, in an evolving political process.[5] The promoters may seek tactical and strategic alliances with sections of the government on common ideological ground (such as national security or support for small farmers).[6] A financial speculator or steel manufacturer may see the railroad as a source of profits, a state official may see a strategic path of communications, and a local official may see the regional benefits of integration into a larger market. Michel Callon's "actor network" of heterogeneous associations captures the political linkages necessary to combine different institutions into supporting a common path of technological advance.[7]

The formation of an alliance is not without its risks: the state may push the technology in ways other than its initial supporters intended (e.g., different priorities for railroad construction, manned over unmanned space flight); the failure of its preferred technology and the neglect of other lines of development may harm the state (e.g., supporting light over heavy water nuclear reactors). An unintended consequence of a close political alliance is the potential loss of support if the faction or government loses power.[8]

A. McDougall, . . . *the Heavens and the Earth: A Political History of the Space Age* (New York: Basic Books 1985).

[4] The local equivalent is a "keeping up with the Joneses" boosterism; see Letty Anderson, "Fire and Disease: The Development of Water Supply Systems in New England, 1870–1900," in Joel A. Tarr and Gabriel Dupuy, eds., *Technology and the Rise of the Networked City in Europe and America* (Philadelphia: Temple University Press, 1988), 149–50.

[5] Bruno Latour, *Science in Action: How to Follow Scientists and Engineers through Society* (Cambridge: Harvard University Press, 1987), 103–44.

[6] Roy Talbert, *FDR's Utopia: Arthur Morgan of the TVA* (Jackson: University of Mississippi Press, 1987).

[7] Michel Callon, "Society in the Making: The Study of Technology as a Tool for Sociological Analysis," in Wiebe E. Bijker, Thomas P. Hughes, and Trevor J. Pinch, eds., *The Social Construction of Technological Systems* (Cambridge: MIT Press, 1987), 92–93.

[8] E. g., the identification of the B-1 as a partisan bomber; see Nick Katz, *Wild Blue Yonder: Money, Politics and the B-1 Bomber* (Princeton: Princeton University Press, 1988).

Old state technologies do not die, but neither do they necessarily fade away. A variety of paths exist: government–industry ties may loosen as interests and priorities change; technologies may become less economically reliant on government support; new technologies may replace the old; or the environment may change so drastically that both the government and the technology lose favor. In the late nineteenth and early twentieth centuries, no self-respecting government could afford not to harness the new industrial technologies to advance its economic and political modernization. The tsarist approach, however, created an environment that hindered industrialization.

To understand the evolution of Russian electric utilities, one must first understand the tsarist state, which feared any activity that threatened its primacy.[9] In comparison with its European counterparts, the Russian government exercised a greater control of the economy by its activities as an authorizer, regulator, producer, and consumer, although it never completely subordinated the economy.[10] Because of their role in the urban infrastructure, utilities operated under more government strictures than did the manufacturing industries. State authority for economic activity diffused into an administrative pluralism[11] in which large bureaucracies battled as they followed uncoordinated and even contradictory policies.[12] Because special interdepartmental committees (which, according to William Fuller, "as any bureaucrat knew could delay the resolution of a conflict for decades"[13]) often failed to coordinate ministerial policies, unified government action proved difficult, if not impossible. To the cost of the economy, the tsarist government did not lend itself to quick decisions.[14]

The main protagonists were the Ministry of Finance, the Ministry of Trade and Industry (MTP, Ministerstvo Torgorlvi i Promyshlennosti, a department of the Ministry of Finance until 1905), and the Ministry

[9] Tim McDaniel, *Autocracy, Capitalism, and Revolution in Russia* (Berkeley: University of California Press, 1988), 23.

[10] Peter W. Gatrell, *The Tsarist Economy, 1850–1917* (New York: St. Martin's Press, 1986), 232; Teodor Shanin, *Russia as a Developing Country. The Roots of Otherness: Russia's Turn of Century*, vol. 1 (London: Macmillan, 1985), 126–30.

[11] Theodore H. Von Laue, *Sergei Witte and the Industrialization of Russia*, 2d ed. (Philadelphia: Lippincott, 1971), 75.

[12] McDaniel, *Autocracy, Capitalism, and Revolution*, 28; Hans Rogger, *Russia in the Age of Modernization and Revolution, 1881–1917* (London: Longman, 1983), 39–47.

[13] William C. Fuller, Jr., *Civil-Military Conflict in Imperial Russia, 1881–1914* (Princeton: Princeton University Press, 1985), 255.

[14] M. C. Kaser, "Russian Entrepreneurship," in *Cambridge Economic History of Europe*, ed. M. M. Postan and H. J. Habakkak (Cambridge: Cambridge University Press, 1978), vol. 7, pt. 2: 416–93.

of Internal Affairs (MVD, Ministerstvo Vnutrikh Del). The Ministry of Finance, especially in the 1890s under Sergei Witte, and the MTP strove to create an institutional infrastructure and political climate conducive to industrial development. The MVD, in addition to overseeing local governments, had the responsibilities of approving the establishment of new industries and developing technical regulations.[15] Industrialization involved other ministries to a lesser extent.[16]

Insofar as the tsarist government supported a technology for industrialization, the railroad was that state technology.[17] Railroads consumed the lion's share of the billions of rubles invested in industrialization.[18] The government constructed, nationalized, and guided the amalgamation of railroads to serve military needs, stimulate the metallurgical and fuel industries, facilitate grain exports to earn hard currency, and create a nationwide transportation system.[19] According to Witte's trickle-down theory, development of this heavy industry would stimulate the development of the more consumer-oriented light industry.

Electrification did not receive the attention given to railroads because it did not appear economically important. Instead, utilities, like other industries, suffered from state overregulation and involvement in almost every area of operations. Utilities dealt with the central government primarily through its local branches. The administrative and financial controls of central ministries over municipal governments essentially constituted a parallel government, which often hurt municipal efforts to improve local conditions and kept local governments politically weak.[20] City *dumas* (legislative councils), elected by only a few, proved quite cautious about approving ventures that required

[15] Von Laue, *Sergei Witte*, 72–75, 92–99; Rogger, *Russia in the Age of Modernization*, 102–5.

[16] E. g., a 1904 attempt to establish a law for hydrostations included petitions to the ministries of finance, internal affairs, justice, communications, state domains, and agriculture: "Deiatelnost Obshchestva," *Zapiski Imperatorskogo russkogo tekhnicheskogo obshchestva (ZIRTO)*, 1904, no. 2: 50.

[17] Roger Portal, "The Industrialization of Russia," in *Cambridge Economic History of Europe*, vol. 7, pt. 1: 814; Shanin, *Russia as a Developing Country*, 128.

[18] Approximately a quarter in 1896–1900 and nearly half in 1900–13; see Gatrell, *Tsarist Economy*, 151–52, 192–94, and J. N. Westwood, *A History of Russian Railways* (London: George Allen and Unwin, 1964), 140.

[19] D. N. Collins, "The Franco-Russian Alliance and Russian Railways, 1891–1914," *Historical Journal* 15, no. 4 (1973): 777–88; Clive Trebilcock, *The Industrialization of the Continental Powers* (London: Longman, 1981), 235–36; Von Laue, *Sergei Witte*, 76, 262–67.

[20] Robert W. Thurston, *Liberal City, Conservative State: Moscow and Russia's Urban Crisis, 1906–1914* (New York: Oxford University Press, 1987), 40.

new technologies or debt, partly because of tight state control of finances and discouragement of local initiative[21] but also because of reluctance to act without expressed authorization from the tsarist government.[22] Resolution of local questions often entailed extensive consultations and negotiations at the state level. For example, deciding which part of the government should pay for replacing kerosene with electric lighting in a St. Petersburg police building took six years and the attention of the state senate.[23] This central dominance of local affairs, coupled with interministerial disputes, hindered the development of the local political initiative essential to introduce and implement new technologies.

To operate, a utility needed several ministerial approvals. The Ministry of Finance had to approve the statutes and capital for every new company. The MVD controlled the regulations governing the construction and operation of electric stations. In 1885, its post and telegraph administration published temporary safety rules for electrical installations and cable networks.[24] In 1890–91, the MVD technical construction committee assumed the responsibility for technical reviews and physical inspections of projects.[25] The MVD did not deal directly with the utility but rather with the *gubernator*, the tsarist-appointed administrator of a city or region.[26] A city government submitted a proposal to the gubernator's committee on rural and urban affairs (Gubernskoe prisutstvie po zemskim i gorodskim delam), which then submitted its recommendation to the gubernator. If he approved, the proposal went to the MVD Main Administration for Municipal Affairs (Glavnoe upravlenie po delam mestnogo khoziaistva) in St. Petersburg. After the Main Administration gave a preliminary approval, the technical construction committee and post and telegraph administration reviewed the project. Requests to seek foreign loans followed the same path but also needed the approval of the Ministry of Finance.[27]

[21] H. Lerche, "State Credit for Town and County Councils," *Russian Review* 1 (1912): 46–48; Thurston, *Liberal City*, 47–49, 54–56, 183.

[22] Of the 140 replies to 700 questionnaires in a 1908 survey, three refused to answer without the permission of their gubernator, a timidity "characteristic of our self-government"; O. G. Flekkel, "VI Vserossiiskii elektrotekhnicheskii sezd," *Gorodskoe delo*, 1911, no. 5: 455.

[23] "Doklad gorodskoi upravy," *Izvestiia S. Peterburgskoi gorodskoi dumy*, 1908, no. 24: 2242–43; 1914, no. 11: 2865–66.

[24] "O vremennykh pravilakh kanalizatsii elektricheskogo toka bolshoi sily i ustroistva provodov i prochikh prisposoblenii dlia elektricheskogo osveshchesniia," *Sbornik rasporiazhenii po glavnomu upravleniiu pochty i telegrafov*, 1886, vol. 1, pt. I: 41–44.

[25] TsGIAL f. 90, op. 1, ed. kh. 466, 6–9.

[26] TsGIAL f. 23, op. 27, ed. kh. 841, 110–12.

[27] Thurston, *Liberal City*, 47.

Reviews were not necessarily rubber stamps. For example, the technical construction committee delayed the construction of the Nizhni-Novgorod municipal station until it made changes, including a stronger foundation in case future demand necessitated turbogenerators instead of vertical engines, a very reasonable demand.[28]

In 1904, a MVD reorganization reduced the authority of the post and telegraph administration to preventing interference with telegraph and telephone lines and increased the purview of the Main Administration for Municipal Affairs. This revision also increased the maximum voltage the gubernator could provisionally approve from the 200 volts set in 1901 to 250 volts.[29] These low voltages meant that nearly every project had to receive MVD approval. In parts, the 1904 rules reprinted verbatim the proposals submitted by the 1st All-Russian Electrotechnical Congress in 1901.[30] Although this can be viewed as an example of the close cooperation between the MVD and the electrical engineering community, it may be more accurate to interpret it as a slow bureaucratic process that demanded three years to produce conservative, technologically outdated regulations. Although it participated in rule making, the electrical engineering community considered the process unwieldy, overly conservative, and a hindrance to the commercial development of new technologies.[31] Efforts to change this process of approval and oversight constantly bogged down over interdepartmental disputes about jurisdiction and policy.

Despite its formidable powers, the state could not simply dictate economic policy but had to negotiate with local governments and industry, as the failure to tax electric energy demonstrates. To pay for the Russo-Japanese war, an interdepartmental commission proposed in 1906, among other measures, a tax on electric energy. Noting that the government taxed kerosene, electricity's main competitor, at approximately 4 kopecks per kilogram, the commission suggested an equivalent tax of 4 kopecks per kilowatt-hour (kWh). Widespread opposition quickly developed from utilities, city dumas, and industrial users, who feared that the tax would cripple the utility industry. The main electrical engineering society, the VI Section of the Imperial Russian Technical Society, simultaneously negotiated details of the tax

[28] "Iz gazet," *Elektrotekhnicheskoe delo*, 1914, no. 5: 21.

[29] TsGIAL f. 23, op. 27, ed. kh. 841, 6–17, 110.

[30] TsGIAL f. 90, op. 1. ed. kh. 471, 7–8.

[31] *Trudy Sedmogo Vserossiiskogo elektrotekhnicheskogo sezda, 1912–1913 gg. v g. Moskve* (St. Petersburg: Postoiannyi Komitet Vserossiiskogo elektrotekhnicheskogo sezda, 1913), 34.

with the Ministry of Finance and filed petitions against it.[32] In 1908, the Ministry of Finance dropped the proposal. War also brought the next proposed tax on electricity, in 1916, but the February revolution intervened before its introduction.[33]

The extensive, albeit distant, state involvement and concomitant slow diffusion of new technologies in Russia were the norm, not the exception: in most areas of public service, the time between proposal in the city duma and final approval was fifteen to twenty years.[34] Compared with other network technologies, electric utilities had an outstanding record of accomplishment: in 1910, 115 cities had utilities but only 40 had sewage systems.[35] Electric utilities spread faster because of the greater availability of foreign technology and financing, a larger customer base, lower construction costs, and the smaller area of coverage needed for profitable service.

The legal framework for electrification was similar to those for other industries in Russia but more restrictive than those in other countries.[36] In Canada and the United States, regulation usually followed rather than preceded new technologies. Although electrification in Europe proceeded under a more regulated regime than in North America, development was also more rapid than in Russia, as the next chapter shows.[37]

The tsarist government retarded the growth of electrification, not by intention but by benign neglect. The state's role was more one of conservative and reluctant authorizer than of entrepreneurial activist. The state neither favored nor disfavored electric utilities; they were

[32] P. P. Dmitrenko, "Ob aktsiz na elektricheskuiu energiiu," *ZIRTO Prilozhenie*, 1908, nos. 9–10: 51–52; "Deiatelnost obshchestva," *ZIRTO*, 1907, no. 11: 459–60; 1908, no. 2: 100–101.

[33] "Khronika," *Elektrichestvo*, 1916, no. 11: 204–5; "Khronika," *Elektrotekhnicheskoe delo*, 1917, no. 5: 14; Alexander M. Michelson et al., *Russian Public Financing during the War* (New Haven: Yale University Press, 1928), 161–65.

[34] Alfred J. Rieber, *Merchants and Entrepreneurs in Imperial Russia* (Chapel Hill: University of North Carolina Press, 1982), 102.

[35] *The Russian Almanac 1919* (London: Eyre and Spottiswoode, 1919), 157.

[36] Rieber, *Merchants and Entrepreneurs*, 97–102, 283.

[37] Christopher Armstrong and H. V. Nelles, *Monopoly's Moment: The Organization and Regulation of Canadian Utilities, 1830–1930* (Philadelphia: Temple University Press, 1986), 130; Thomas P. Hughes, *Networks of Power* (Baltimore: Johns Hopkins University Press, 1983), 58–61, 71–72; Leslie Hannah, *Electricity before Nationalization* (Baltimore: Johns Hopkins University Press, 1979), 5–8; Brian Bowers, *A History of Electric Light and Power* (London: Peter Peregrinus, 1982), 152–61; John McKay, "Comparative Perspectives on Transit in Europe and the United States, 1850–1914," in Tarr and Dupuy, eds., *Rise of the Networked City*, 5–20.

simply one of many regulated activities. The major exception to the state lack of interest in electrotechnology came from the military.

The Role of the Military

IN THE 1870s in Russia, the electric light left the laboratory and ventured into the public domain. A distinguishing feature of this transition was the leading role of the army and navy. The military was similarly involved elsewhere, but only in Russia was it so important.[38] Over a decade before the first utilities came into being, the military provided the initial base and market for electrification, and it retained this leading position through the 1890s.

Unlike the civilian ministries, the military actively nurtured electrotechnology in a protective, fertile environment until the new technology could survive in the harsher civilian sphere. Compared with Europe, the Russian civilian economy was weakly developed and less industrialized. The army and navy commanded the resources to finance and develop new technologies, they had specific needs, and economic feasibility was subordinated to national security. And the consequences of failure were not as severe for military entrepreneurs. In such circumstances, the military's large role is understandable.

The military has been influential in the development of science, technology, and industry worldwide. Its most vital activities have been educating and hiring technical personnel, serving as an initial customer, and promoting domestic and international technology transfer. Other important ways of promoting new technologies include funding and conducting research and development, fostering domestic industry, and creating standards. In Russia, the army and navy engaged in all these activities, serving as a Gerschenkronian state substitute for the industrial development lacking in the backward civilian economy.[39]

The Russian military found many uses for electrotechnology. Electricity could detonate torpedoes and explosive mines, turn night into day outside fortresses, safely illuminate factories, transmit informa-

---

[38] Jonathan Coopersmith, "Electrification and the Military, 1870–1900," paper presented to the British Society for the History of Science Conference on Society and War, London, 1989.

[39] Alexander Gerschenkron, *Economic Backwardness in Historical Perspective* (Cambridge: Harvard University Press, 1966), 123–24; Trebilcock, *Industrialization of the Continental Powers*, 222.

tion, run clocks, and provide power. Although the army and navy conducted separate research and testing programs and deployed different equipment, they cooperated formally and informally via the exchange of information and personnel. Officers worked on the advisory committees of their brother service and assisted in testing, installation, and education.[40] These cross-service links helped spread electrotechnology within the military.

Army interest in electricity began in the late 1860s. The Main Artillery Administration (GAU, Glavnoe artilleriiskoe upravlenie) dominated army electrical engineering through the 1880s. The GAU was much more than simply the artillery arm of the army. Its troops, trained in GAU schools and academies, staffed fortresses equipped with weapons built and tested by its workshops, factories, and arsenals. GAU factories introduced new ideas and technologies, like the Harpers Ferry and Springfield armories did in the United States.[41] For example, Col. Vasilii F. Petrushevskii established an instruments section at the St. Petersburg cartridge factory that standardized mechanics' instruments and training in the mid-1870s.[42] Petrushevskii's activities typify the standardized testing and hierarchical control that characterize military technology.[43] Besides formal research and testing, unofficial research occurred at GAU installations at the discretion of the commander, but its very informality precludes an accurate assessment of its pervasiveness and importance.[44] Certainly such research supports the concept of Russian industrial fiefdoms in which the director had a great deal of leeway in managing his operations. Such activities indicate supportive environments for scientists and engineers.

At the peak of the GAU's technological investment stood the artillery committee, or Artkom. Established in 1869, Artkom succeeded the technical committee in directing GAU's technical priorities, allocating resources, and appraising Russian and foreign research. In

[40] "Mikhail Matpevich Boreskov," *Elektrotekhnik*, 1897–98, no. 8: 495–96; P. Berkhman, *Sudovye miny: Rukovodstvo dlia slushatelei minnogo ofitserskogo klassa* (St. Petersburg, n.d.), 1; TsGVIA f. 506, op. 1, d. 409, 537, 547–48, 552.

[41] Merritt Roe Smith, *Harpers Ferry Armory and the New Technology: The Challenge of Change* (Ithaca: Cornell University Press, 1977).

[42] "General-Leitenant Petrushevskii (nekrolog)," *Russkii invalid*, 1 May 1891, 4; "Petrushevskii," *Entsiklopediia voennykh i morskikh nauk* (St. Petersburg, 1891), 5: 628; A. Ia. Averbukh, *Vasilii Formich Petrushevskii* (Moscow: Gosenergizdat, 1967), 14.

[43] "Introduction," in Merritt Roe Smith, ed., *Military Enterprise and Technological Change* (Cambridge: MIT Press, 1985), 17–21.

[44] "N. M. Alekseev," *Elektrichestvo*, 1903, no. 4: 48–49; A. A. Chekanov and B. N. Rzhonsnitskii, *Mikhail Andreevich Shatelen, 1866–1957* (Moscow: Nauka, 1972), 12.

an example of stimulation by knowledge about work elsewhere, the GAU technical committee in July 1868, sparked by a Prussian article about harbor lighting in the American civil war, asked Col. Petrushevskii to study searchlights for fortress defense.[45] Petrushevskii was the embodiment of the Russian scientist-soldier. Educated in the military schools where he later taught, he conducted research on electric mines and lighting and invented an artillery rangefinder. Petrushevskii was a consulting member of Artkom until 1881, when he was promoted to lieutenant-general and became a permanent member. He founded and headed its electrotechnical department in 1886 until his death in 1891.[46]

Petrushevskii tested Drummond lamps, magnesium lights, battery-powered lamps, and arc lights. With the exception of the last, these systems were mature technologies. Thomas Drummond, for example, invented his "limelight" in 1826. Petrushevskii's tests, completed in 1870 significantly over budget and schedule, demonstrated the "full advantage of electric light," but he continued to study recent European equipment.[47] The tests were a model of how to judge new technologies, with frequent trips abroad to inspect the latest developments and inspectors to assure that factories sent functional equipment. Petrushevskii's problem was when to halt testing and actually install a specific system, knowing that better systems would soon appear. In this case of the perennial conflict between developers and users, the decision came from his superiors, whose interest was not the most advanced technology but the best defense of their fortresses.[48]

GAU involvement with electricity created a career pattern for its technical officers similar to their modern counterparts, with management in the factory and office as important as command of troops. Less common, but not unusual, were assignments to other parts of the government to install electric lighting.[49] Many officers taught, lec-

---

[45] TsGVIA f. 506, op. 1, d. 409, 3–4.

[46] "General-Leitenant Petrushevskii," 3–4; Averbukh, *Petrushevskii*.

[47] TsGVIA f. 506, op. 1, d. 409, 395–96, 424.

[48] For more information, see Jonathan Coopersmith, "The Role of the Military in the Electrification of Russia, 1870–1890," in E. Mendelsohn, M. R. Smith, and P. Weingart, eds., *Science, Technology and the Military* (Dordrecht: Kluwer Academic Publishers, 1988), 12: 291–305.

[49] E. g., A. I. Smirnov spent two decades working for the Ministry of the Court; see Ia. I. Senchenko, "Vydaiushchiisia elektrotekhnik Aleksandr Ivanovich Smirnov," *Trudy Instituta po istorii estestvoznaniia i tekhniki*, 1962, no. 44: 171–78.

tured, or wrote manuals as part of their duties. Indeed, the GAU published some of the first Russian books on electrotechnology.[50]

Navy activity paralleled the army's. The navy began experimenting with French electric searchlights in 1869 in the Baltic Sea, and tests continued through 1873.[51] In 1874, the Russian navy was the world's first to switch from the Alliance generator to the new, more powerful Gramme generator, a sign of technical leadership and financial backing.[52] The navy established the Mine Officer Class in October 1874 at Kronstadt near St. Petersburg as its center for electrotechnical training, testing, and research.[53] The navy had a large investment in electrotechnology, including an explosives factory, manufacturing facilities, and repair shops—in effect, a self-contained industrial complex. The officers and men formed one of the few competent pools of electrical workers in Russia. Besides training, the Mine Officer Class tested equipment and electrified government buildings and events.[54] The 1881 coronation of Aleksandr III featured a massive display of searchlights and Edison incandescent lights in the Kremlin by the Mine Officer Class assisted by English engineers.[55]

The military provided opportunities for civilians, often on groundbreaking projects. The most prominent example of a civilian expert working for the military was Vladimir N. Chikolev. Officially a low-level GAU clerk, Chikolev proved as important as Petrushevskii in guiding the army's adoption of electricity. He was a dynamic entrepreneur in both military and civilian spheres, albeit more successful in the former because of the military's friendlier environment for elec-

[50] V. N. Chikolev, *Elektricheskoe osveshchenie v primenenii k zhizni i voennomu iskusstvu* (St. Petersburg: F. Pavlenkov, 1885), and *Lektsii po elektrotekhnike* (St. Petersburg: Artilleriiskii zhurnal, 1887).

[51] "Otchet predsedatelia uchenogo otdeleniia morskogo tekhnicheskogo komiteta i komiteta morskikh uchebnykh zavedenii za 1871," *Morskoi sbornik*, September 1872, no. 1: 9; "Otchet predsedatelia uchenogo otdeleniia morskogo tekhnicheskkogo komiteta i komiteta morskikh uchebnykh zavedenii za 1872," *Morskoi sbornik*, September 1874, no. 9: 11–12.

[52] Rondolphe van Wetter, *L'Eclairage électrique à la guerre* (Paris: G. Carre, 1889), 82; Em. Alglave and J. Boulard, *The Electric Light: Its History, Production and Applications* (New York: Appleton, 1884), 393.

[53] "Polozhenie o Minnom ofitserskom klasse i o minnoi shkole dlia nizhnikh chinov," *Morskoi sbornik*, March 1875, no. 3: 25.

[54] E. g., Edison, Swan, and Maxim incandescent lights in 1882–83; *Materialy k istorii Minnogo ofitserskogo klassa i shkoly* (St. Petersburg: Minnyi ofitserskii klass, 1899), 95.

[55] "Notes," *Telegraphic Journal and Electrical Review*, 24 February 1883, 334, and 30 June 1883, 538; V. Iu. Gorianov, "Ie. P. Tveritinov—osnovopolozhnik sudovoi elektrotekhniki v Rossii," *Elektrichestvo* 1960, no. 12: 78–81.

trification. A graduate of military schools and an external student at Moscow University, he worked in Moscow as a laboratory assistant and for Pavel N. Jablochkov's electric light company. After a bad investment depleted his resources, Chikolev moved to St. Petersburg in 1876. He served as the initial editor of *Elektrichestvo*, the first Russian electrotechnical journal. His company, Elektrotekhnik, attempted to light Nevskii Prospekt in St. Petersburg in 1880. Although this financial failure was bought out by Siemens and Halske, he did install some electric street lights in Moscow in 1883.[56] Chikolev also published a novel about electricity, in 1895.[57]

The GAU hired Chikolev as a clerk in 1877, beginning an association that lasted until his death in 1898. Working closely with Petrushevskii, Chikolev organized electric lighting systems for fortresses, reviewed research proposals, developed searchlights, tested new equipment, and traveled abroad for the GAU.[58] He nearly saw frontline duty with a mobile searchlight unit in the 1877–78 Russo-Turkish war, but typhus intervened.

Until the diffusion of utilities in the 1890s, the army and navy provided the major markets and support for electric lighting. The military strengthened the Russian scientific and technical infrastructure by supporting research, education, technical societies, foreign trips, and prize competitions. It also tested materials and equipment, which assisted the development of a domestic industry and aided standardization.[59] Possibly the most important contribution was thousands of engineers and technicians educated in military schools and academies, far more than in all civilian schools.[60] By creating this infrastructure and market in the 1870s and 1880s, the military provided the underpinnings of later civilian electrification. Thomas J. Misa's description of the development of the transistor in America applies equally well

[56] "V. N. Chikolev," *Elektrotekhnik*, 1897–98, no. 8: 497–502; "Vladimir Nikolaevich Chikolev," *Entsiklopedicheskii slovar* (St. Petersburg: Brogaus-Efron, 1903), vol. 76: 826–27; N. A. Shotsin, "Vladimir Nikolaevich Chikolev," *Elektrichestvo*, 1945, no. 8: 7–12; I. D. Artamonov, "V. N. Chikolev—voennyi elektrotekhnik," ibid., 13–16; V. V. Zapolskaia, "Iz vospominanii V. V. Zapolskoi o V. N. Chikoleve," *Elektrichestvo*, 1948, no. 6: 77–79.

[57] *Ne byl, no i ne vydumka—electricheskii razskaz* (St. Petersburg: Babkin, 1895), cited in Richard Stites, *Revolutionary Dreams: Utopian Vision and Experimental Life in the Russian Revolution* (New York: Oxford University Press, 1989), 30.

[58] E. g., TsGVIA f. 506, op. 1, d. 437, 42, 80, 88, 154, 191, 198, 247, 293.

[59] Voennoe Ministerstvo, *Vsepoddanneishii otchet Voennogo Ministerstva za 1881 god* (St. Petersburg: Gogenfelgen, 1883), 18–19; *Vsepoddanneishii otchet Voennogo Ministerstva za 1892 god* (St. Petersburg: Gogenfelgen, 1894), 34.

[60] *Materialy k istorii Minnogo ofitserskogo klassa*, 257–58.

to electric lighting in Russia a century earlier: "Military sponsorship helped shield the new technology from undue criticism and economic constraint and also provided the necessary potential to push it through the development stage to commercialization."[61] The introduction and promotion of electrotechnology saw the military sector paralleling Western activities and the civilian sector lagging, a frequent pattern in Russian history. Without the Russian military, electrotechnology would have developed even more slowly, more expensively, and with more foreign involvement than it did.

Technical Societies

SCIENTIFIC AND TECHNICAL societies have played major roles in the creation, diffusion, and application of knowledge.[62] They have played a no less important role in the development, professionalization, and political activities of the knowledge holders themselves. As technical knowledge became more important for the industrializing economy, so did the technical societies.[63]

Russian engineers founded their first technical society, the Imperial Russian Technical Society (IRTO, Imperialskoe Russkoe Tekhnicheskoe Obshchestvo), in 1866. Aided by government funding, the IRTO was oriented toward industry and the military, with sections for chemical production and metallurgy, mechanics and machine construction, construction and mining, and naval and military technology.[64] An umbrella organization, the IRTO expanded to fifteen sections and forty local branches by World War I. Engineers founded more than forty-five other technical societies.[65] Increasingly located outside St. Petersburg after 1905, these societies reflected the geo-

[61] Thomas J. Misa, "Military Needs, Commercial Realities, and the Development of the Transistor, 1948–1958," in Smith, ed., *Military Enterprise*, 255.

[62] "Scientific Institutions," in *Dictionary of the History of Science* (Princeton: Princeton University Press, 1984), 377–78.

[63] Edwin T. Layton, Jr., *The Revolt of the Engineers* (Baltimore: Johns Hopkins University Press, 1986).

[64] James H. Swanson, "The Bolshevization of Scientific Societies in the Soviet Union" (Ph.D. diss., Indiana University, 1968), 21.

[65] N. G. Filippov, *Nauchno-tekhnicheskie obshchestva Rossii (1866–1917)* (Moscow: Moskovskii gosudarstvennyi istoriko-arkhivnyi institut, 1975), 32–33, 35. At least forty-five: Filippov omits the Russian Electrical Society, founded in 1900 at the St. Petersburg Electrotechnical Institute; see M. A. Shatelen, "Russkoe elektricheskoe obshchestvo," *Elektrichestvo*, 1900, nos. 22–24: 351–52.

graphic spread of industry and the growing professionalization of the engineering community.[66]

The first electrotechnical organization grew from the IRTO and scientific societies in St. Petersburg and Moscow.[67] Late in 1879, a group of engineers petitioned the IRTO to form a new section. On 30 January 1880, fifty-six people attended the first meeting of the new VI (electrotechnical) Section in St. Petersburg.[68] Like other sections, the VI Section consulted on projects, developed official standards, petitioned and worked with the government, collected information, advanced its members' prestige, and popularized electrification.

Continuing the IRTO orientation, a military presence dominated the early years of the VI Section: Gen. F. K. Velichko was president, candidate-president Pavel N. Jablochkov's firm dealt mainly with the navy, and the military employed at least three of the nine permanent members. This military involvement remained strong through the 1880s. Of the eighty-two active members in 1885, half worked in or for the military.[69] A sample of twenty-five active members in 1889 found eleven military employees, a slight drop by percentage.[70]

Military support did not benefit only the VI Section. The Electrotechnical Society, established in 1892, initially met in the St. Petersburg Naval Museum and received other navy support.[71] Electrical exhibits benefited from War Ministry awards and exhibits.[72]

In 1880, the VI Section published the first IRTO section journal, *Elektrichestvo* (Electricity). Despite financial and editorial struggles, *Elektrichestvo* continued to publish until 1918.[73] Other journals appearing in the 1890s focused on more practical applications—*Elektrotekhnicheskii vestnik* (Electrotechnical Herald) and *Elektrotekhnik* (Electrotechnician)—or were directed to technicians rather than engineers—

---

[66] After 1905, sixteen of twenty-nine new societies formed outside the capital, compared with six of eighteen in the four decades before 1905; see Filippov, *Nauchno-tekhnicheskie obshchestva*, 206–13.

[67] Lev D. Belkind, *Pavel Nikolaevich Jablochkov* (Moscow: Izdatelstvo Akademii SSSR, 1962), 57–67.

[68] "Kratkii obzor deiatelnosti," *Elektrichestvo*, 1880, no. 1: 2.

[69] Forty-two members in military or military-related occupations, twenty-three nongovernment, ten in civil government, six academic, and one unknown; see "Sostav IRTO," *ZIRTO*, 1885, no. 2: 8–33.

[70] "Lichnyi sostav IRTO," *ZIRTO*, 1890, no. 7: 1–55.

[71] "Elektrotekhnicheskoe obshchestvo," *Elektrotekhnicheskii vestnik*, 1894, no. 1: 5.

[72] E. g., "Raznye izvestiia," *Elektrichestvo*, 1888, no. 15: 142; "Uspeki v elektrotekhnike,' *Elektrichestvo*, 1893, no. 1: 2.

[73] M. A. Shatelen, "'Elektrichestvo' (1880–1930)," *Elektrichestvo*, 1930 Jubilee Issue, 3–4; A. V. Netushil and Ia. A. Sheibert, "Osnovanie zhurnala 'Elektrichestvo' i pervykh dvadtsat let ego deiatelnosti," *Elektrichestvo*, 1979, no. 7: 1–11.

Elektricheskoe delo (Electrical Affairs)—but *Elektrichestvo* remained the preeminent Russian electrical journal.

The VI Section provided invaluable technical and economic expertise to city administrations. Its commissions studied a city's technical demands, judged proposals, calculated operating costs, and worked out consumption, system efficiency, and the best equipment.[74] The numerous requests for assistance and guidance literally buried the section as utilities spread after 1900.[75] From 1899 to 1914, the section handled more than fifty requests from cities and towns; that is, it assisted one-third of all electrified cities.[76]

Like electrotechnical societies elsewhere, the VI Section worked with state ministries on issues ranging from standards to siting.[77] Members served on government panels, formed committees to handle government requests, and published standards in *Elektrichestvo*. The VI Section and, after 1900, the Permanent Committee of the All-Russian Electrotechnical Congresses routinely petitioned the MVD for changes in laws and regulations. The section usually worked with the MVD and Ministry of Trade and Industry, but it also dealt with other ministries on specific issues, such as the Ministry of Finance's proposed tax on electric energy.[78] Although it worked well with the MVD and MTP, the VI Section's influence was fairly weak, for it had no active constituency inside the government until World War I.

Membership was small for Russia's leading electrotechnical society. The section contained approximately 140 members in 1891 (90 percent in St. Petersburg), 156 active members in 1906, 196 active members in 1908, and 243 active members in 1910.[79] The latter was only one-third the average attendance at the All-Russian Electrotechnical Congress

---

[74] "Zaklucheniye Komissii po rassmotreniiu tekhnicheskikh zadanii na ustroistvo elektricheskogo osveshcheniia v g. Nizhnem-Novgorode i po rassmotreniiu predstavlennykh proektov," *ZIRTO*, 1907, no. 6: 329.

[75] Filippov, *Nauchno-tekhnicheskie obshchestva*, 132.

[76] See TsGIAL f. 90, op. 1, ed. kh. 480–82, and the regular "Deistviia Obshchestva zhurnala zavedenii VI-ogo otdela" section in *ZIRTO*. See also, Filippov, *Nauchno-tekhnicheskie obshchestva*, 132.

[77] E. g., the German Verband Deutscher Elektrotechniker; see "The German Electrotechnical Societies," *Electrical World*, 2 February 1911, 290.

[78] The archival records of the VI Section are rich with these communications (e. g., TsGIAL f. 90, op. 1, ed. kh. 456–58, 466, 471, 480–82). See also, "Sobraniia chlenov VI otdela IRTO," *Elektrichestvo*, 1901, nos. 11–12: 176, and "Otchet o deiatelnosti VI otdela," *Elektrichestvo*, 1906, nos. 11–12: 160.

[79] For 1891, see TsGIAL f. 90, op. 1, ed. kh. 458, 68–69; for 1906 and 1908, see "Deiatelnost obshchestv," *ZIRTO*, 1908, nos. 6–7: 280–83; for 1910, see "Otchet o deiatelnosti IRTO v 1910 godu," *ZIRTO*, 1911, nos. 6–7: 247. Categories also existed for honorary and inactive members.

or one-half the membership of the Moscow-based Society of Electro-technicians that year.[80] By comparison, the British Institution of Electrical Engineers had 4,010 members in 1901, the German Verband Deutscher Elektrotechniker had 4,653 members in 1910, and the American Institute of Electrical Engineers had 7,100 members in 1910.[81]

The VI Section and *Elektrichestvo* remained unique until the creation of new electrical journals in the 1890s to serve the growing number of electrical engineers. The economic boom after the 1905–6 revolution further expanded the number and size of professional societies. The spread and geographic concentration of these societies reflects the slow diffusion of electrification. Of the seven prewar electrotechnical societies, only two existed before 1900 and three began in 1909. St. Petersburg housed four societies; Moscow, Kharkov, and Kiev each contained one.[82]

Electrotechnical societies cooperated more than they competed; multiple membership was not uncommon. They jointly sponsored the biannual All-Russian Electrotechnical Congress, a united, albeit weak, voice of the electrical engineering community. Between 1899 and 1913, congresses met seven times in St. Petersburg, Moscow, and Kiev and drew an average of 600 people.[83] The war caused the cancellation of the eighth meeting, planned for Kharkov in 1915. The congresses promoted professionalism, passed resolutions, and served as clearinghouses for the latest technical information.[84]

The Electrical Engineers

THE VI SECTION and other societies played a vital role in the creation and diffusion of electrification, especially in promoting and providing technical knowledge and skills. Societies, however, ultimately de-

[80] "Deiatelnost Obshchestva elektrotekhnikov v Moskve," *Elektrichestvo*, 1914, no. 9: 291.

[81] British data produced by Geoffrey Tweedale for W. J. Reader's *A History of the Institution of Electrical Engineers* (London: IEE, 1987); "The German Electrotechnical Societies," *Electrical World*, 2 February 1911, 287; *AIEE Yearbook* (New York: AIEE, 1914), 19.

[82] Filippov, *Nauchno-tekhnicheskie obshchestva*, 206–13; M. A. Shatelen, "Russkoe elektricheskoe obshchestvo," *Elektrichestvo*, 1900, nos. 22–24: 351–52.

[83] B. S. Sotin and L. G. Davydova, "Russkie elektrotekhnicheskie sezdy," *Trudy Instituta istorii estestvoznaniia i tekhniki* 26 (1959): 6–41.

[84] E. g., the original draft of the contract between the 1886 Company and the Bogorod local government stated that MVD rules would guide the placement of transmission lines, but that information from the fourth and fifth congresses would guide operations; TsGIAMO f. 722, op. 1, ed. kh. 876, 2.

pended on their individual members, the electrical engineers. Engineers provided both the skilled personnel to construct, operate, and expand Russian utilities and a firm link with the international electrotechnical community. These men—and very few women[85]—formed the technical societies, educated and trained their successors, advised cities and, with less success, the national government, and proposed plans for the social and industrial transformation of Russia.

Two groups dominated the electrotechnical community through the early years of Soviet power. St. Petersburg housed one group in educational institutes and firms. The Moscow section of the 1886 Company, the country's largest utility, and Elektroperedacha, Russia's first regional station, housed the second group. A third group of Moscow academics in the heat committee contributed greatly to electrification planning after 1914 but never assumed leadership. These groups controlled the VI Section and other professional activities. During the tsarist era, these engineers were academics or high-level managers for German utilities. Wartime participation in state, city, and Central War Industries Committee activities introduced electrical engineers into the country's leadership circles. After the February revolution, these engineers built a base in the government. After the October revolution, they took charge of developing and implementing state electrification plans.

The VI Section served as an institutional focus for St. Petersburg electrical engineers, who worked for a range of employers. Academia contributed the most prominent engineers, followed by manufacturers, utilities, and, before 1895, the military. Among the academics were professors Mikhail A. Shatelen, who tried to professionalize electrical education and the VI Section; P. D. Voinarovskii, the director of the Electrotechnical Institute after 1906; Aleksandr V. Vulf, a railroad electrification advocate; and Piotr S. Osadchii, who led electrical engineers into close cooperation with the provisional government in 1917. Leonid B. Krasin was the most notable electrical engineer in industrial management. The military figures included Gen. F. K. Velichko, the first president of the VI Section, the inventor Jablochkov, and Chikolev, electrotechnology's Renaissance man.

The utility-based Moscow group stood out as a proving ground for

---

[85] The St. Petersburg Women's Technical Institute, established in 1906, had graduated only fifty female engineers by 1916; see Richard Stites, *The Women's Liberation Movement in Russia: Feminism, Nihilism, and Bolshevism, 1860–1930* (Princeton: Princeton University Press, 1978), 176. See also V. M. Buzinova-Dybovskaia, "Pervye zhenskie politechnicheskie kursy," *Elektrichestvo*, 1970, no. 7: 91–92.

Russian manager-engineers and for its Bolshevik electrical engineers, one of the few such prerevolutionary clusters. The 1886 Company's Moscow section consciously switched from German to Russian managers, engineers, and technicians after 1900. Its summer program for students attracted young men from all the country's technical institutes and allowed the 1886 Company to select and groom promising future engineers.[86] The Moscow section of the 1886 Company and Elektroperedacha became a haven for Bolsheviks, who held major positions of responsibility before 1917, including Gleb M. Krzhizhanovskii, Robert E. Klasson, Aleksandr V. Vinter, Ivan I. Radchenko, Piotr G. Smidovich, and V. Z. Esin.[87]

The professional and political links among the Bolshevik engineers began in their student days at the St. Petersburg Technological Institute, where in 1890 Klasson founded the first study group to introduce successfully Marxism to workers. The original group included Nadezhda Krupskaia, Lenin's future wife, and Stephan I. Radchenko, "perhaps the first truly professional *apparatchik*." An excellent example of an "old-boy network," the graduates of the St. Petersburg Technological Institute, the center of Russian electrotechnical education, formed a "kind of electrician's mafia" which "enjoyed a certain immunity from prosecution because of the desperate need in a rapidly industrializing economy for native technology."[88]

After working in Germany with Mikhail O. Dolivo-Dobrovolsky on long-distance transmission and studying Marxism, Klasson directed Russia's first 3-phase AC project in 1895 at the GAU Okhtensk gunpowder factory; this was the last major example of military leadership in electrotechnology.[89] Two decades later, Klasson, Vinter, and Ivan I. Radchenko, brother of Stephan, led the prerevolutionary and Soviet efforts to utilize peat and brown coal. Krzhizhanovskii, future head of GOELRO and Gosplan, directed the 1886 Company cable network in

[86] TsGANKh f. 9508, op. 1, ed. kh. 14, 4. TsGIAMO f. 722, op. 1, d. 602 contains scores of summer job applications.

[87] Mark O. Kamenetskii, *Robert Eduardovich Klasson* (Moscow: Gosenergoizdat, 1963), 78–79; Gleb V. Lipenskii, *Moskovskaia energeticheskaia* (Moscow: Moskovskii rabochii, 1976), 19–23, 27; Vladimir Kartsev, *Krzhizhanovskii* (Moscow: Molodaia gvardiia, 1980), 226–27; Alek G. Cummins, "The Road to NEP, the State Commission for the Electrification of Russia (GOELRO): A Study in Technology, Mobilization and Economic Planning" (Ph.D. diss., University of Maryland, 1988), 23.

[88] According to James H. Billington, *Fire in the Minds of Men: Origins of the Revolutionary Faith* (New York: Basic Books, 1980), 448, 453–55.

[89] R. E. Klasson, "Elektricheskaia peredacha sily trekhfaznymi tokami na Okhtinskikh porokhovykh zavodakh bliz Peterburga," *Elektrichestvo*, 1897, no. 19: 257–67; Kamenetskii, *Klasson*, 13–15; Billington, *Fire in the Minds*, 448.

Moscow. At a lower level of the company worked Smidovich, who had been expelled from Moscow University in 1895 for political agitation and completed his education in electrical engineering in Paris.[90]

Another member of Klasson's study group was Krasin, who served as a director of Siemens and Halske, the main electrotechnical manufacturer in Russia and a major supplier for the 1886 Company. While constructing and operating the electric utility in Baku in 1900–4, Krasin used his position to hire and protect other Bolsheviks. Klasson, himself in internal exile for his political activities, had given Krasin the Baku position, which Krasin used to construct and operate an illegal printing plant. At one point, Krasin purchased printing equipment from a 2,000 ruble city loan intended for utility expenses. During the war, Krasin worked for the Central War Industries Committee to organize trading resources, a task he continued under Soviet rule.[91]

The evolution of the electrical engineering community reflected the development of electrotechnology in Russia. St. Petersburg, with its preponderance of educational, military, and industrial facilities, housed the leadership of the electrical engineering community, though Moscow increasingly took the technological and political lead after 1910.

Inventions

ONE GLARING EXCEPTION to the accomplishments of the Russian electrical engineering community was invention—only the first step in the larger process of translating an idea into a commercial success. A good idea is not enough; its creator must endow it with the social and economic characteristics it needs for survival.[92] The paucity of commercially successful inventors is a striking aspect of Russian electrification and indicative of the societal and economic weaknesses that hindered its development.

Russian engineers and scientists were not passive recipients of for-

[90] Vasilii Iu. Steklov, *Lenin i elektrifikatsiia*, 3d ed. (Moscow: Nauka, 1975), 169.

[91] *Who Was Who in the Soviet Union* (Metuchen: Scarecrow Press, 1972), 311; Michael Glenny, "Leonid Krasin, the Years before 1917: An Outline," *Soviet Studies* 22 (1970), 194–95; Billington, *Fire in the Minds*, 461; Robert W. Tolf, *The Russian Rockefellers: The Saga of the Nobel Family and the Russian Oil Industry* (Stanford: Hoover Institution Press, 1976), 154; Lubova Krassin, *Leonid Krassin, His Life and Work* (London: Skeffington and Son, 1929), 41.

[92] See, e. g., Thomas P. Hughes, "The Evolution of Large Technological Systems," in Bijker, Hughes, Pinch, eds., *Social Construction of Technological Systems*, 63.

eign technologies; they invented and developed their own equipment too. A glance through the pages of *Elektrichestvo* quickly dispels any notion of a lack of creativity. Many ideas brought to fruition in the West, such as incandescent lighting, had Russian counterparts in conception and experimentation although not in transfer and production. Yet only three Russian inventors received national and international recognition in the late nineteenth century: Aleksandr N. Lodygin, Jablochkov, and Dolivo-Dobrovolsky.[93] Why did invention not translate into success in innovation and application? Responsibility falls on two intertwined causes: a systemic failure of the Russian economic and social environment to support and foster domestic inventions, and technological prematurity, the development of an idea before its supporting materials and components attain technological and economic feasibility.

Invention does not occur in a vacuum. The frequency of simultaneous discovery and invention illustrates the extent to which separate inventors share a common world of interests, materials, equipment, financing, and ideas.[94] Thomas Edison has been widely recognized and promoted as the inventor of the incandescent light,[95] but many others invested time and money in the quest for a commercially viable incandescent light.[96] A few professional inventors, such as Elmer Sperry, successfully combined good ideas, financial backing, and customer support;[97] most, however, failed. Failure is a normal outcome in technological development; success, the exception. Perhaps Russia was unexceptional and Europe the aberration. But what made Russia so unexceptional?

The inventor did not find Russia hospitable. Although research facilities existed in military and civilian educational institutes, financial

[93] A close contender is Achilles de Khotinsky, a former naval officer who participated in the early searchlight experiments and manufactured light bulbs in Russia and the West in the 1880s; see "Achilles de Khotinsky," *National Cyclopaedia of American Biography* (New York: J. T. White, 1936), 25, 63–64; A. Heerding, *The History of N. V. Philips' Gloeilampenfabrieken: The Origin of the Dutch Incandescent Lamp Industry*, vol. 1 (Cambridge: Cambridge University Press, 1986), 139–40, 148.

[94] Robert K. Merton, "Singletons and Multiples in Science," in Norman W. Storer, ed., *The Sociology of Science* (Chicago: University of Chicago Press, 1973), 343–82.

[95] Wyn Wachhorst, *Thomas Alva Edison: An American Myth* (Cambridge: MIT Press, 1981).

[96] E. g., Moses G. Farmer, Hiram S. Maxim, St. George Lane-Fox, and Joseph W. Swan; see Arthur A. Bright, Jr., *The Electric Lamp Industry: Technological Change and Economic Development from 1800 to 1947* (New York: Macmillan, 1949), 42–55.

[97] Thomas P. Hughes, *Elmer Sperry: Inventor and Engineer* (Baltimore: Johns Hopkins University Press, 1971).

support proved hard to obtain, manufacturing was difficult, and weak sales diminished profits. Furthermore, the quality, robustness, and suitability of the marketed products played a critical role. Technological prematurity, facing weaknesses in materials, equipment, components, and theoretical approaches, can keep a good idea from fruition.[98] Whereas Lodygin's incandescent lamp failed technically and commercially in Russia in the early 1870s, Jablochkov's arc lamp succeeded in Europe in the mid-1870s partly because he took advantage of advances in supporting components in the intervening three years. The two lamps offer a study in contrasts.

Lodygin's incandescent light bulb, developed in 1872, received the Lomonosov Prize from the Academy of Sciences in 1874 despite bad design, an inadequate vacuum, and poor filaments.[99] Lodygin's lamp underwent constant modification by the inventor and his senior mechanic, Vasilii F. Didrikhson. They tested different materials and designs to increase the duration of burning, the brightness of the light, and the strength of the vacuum.[100] Lodygin worked without benefit of the army's Volkovo field test facility, despite a request to use the installation.[101] Military interest in searchlights did not extend to an outsider working on a smaller light. Lodygin formed a company to manufacture and market an improved version, but a light "more appropriate for laboratory tests and lectures than continual lighting" ruined his first financial backer.[102] The major problem that plagued him, as many other unsuccessful inventors, was the disintegration of the carbon filament.[103] In 1875, Lodygin had to work at the St. Petersburg arsenal as a metalworker, despite the efforts of another investor, banker Stanislav V. Konn. Konn marketed an improved version under his name with a Gramme generator, but he died in late 1876.[104] Lodygin's company, unable to find further support, withered away.[105]

---

[98] E. g., the theory of forward-swept wings preceded the availability of the needed composite materials and computers by four decades; see Gadi Kaplan, "The X-29: Is It Coming or Going?" *Spectrum*, June 1985, 54–60.

[99] Liudmila N. Zhukova, *Lodygin* (Moscow: Molodaia gvardiia, 1983), 117–19, 137–45.

[100] E. O. Bukhgeim, "K istorii vozniknoveniia elektricheskogo osveshcheniia," *Pochtovo-telegraficheskii zhurnal*, 1900, no. 2: 158–63; Ia. I. Kovalskii, ed., *Ocherk rabot russkikh po elektrotekhnike s 1800 po 1900 god* (St. Petersburg; 1900), 35–41; "Vasilii Fedorovich Didrikhson," *Elektrichestvo*, 1930, no. 14: 615.

[101] TsGVIA f. 506, op. 1, ed. kh. 419, 770, and 774.

[102] Kovalskii, *Ocherk rabot russkikh*, 38.

[103] Alglave and Boulard, *Electric Light*, 119–21.

[104] Van Wetter, *L'Eclairage électrique*, 82.

[105] The company ceased paying its gold duty after 1876; TsGIAL f. 1287, op. 7, ed. kh. 2618, 3.

Financial speculation, a recurring problem of start-up firms, may have aided the firm's demise.[106]

From 1878 to 1884, Lodygin worked for Jablochkov's company. Beginning in 1881, Jablochkov's firm manufactured Lodygin's Russian lamp until it was overwhelmed by imported Edison incandescent lamps.[107] For the next two decades, Lodygin worked on electric lighting in France and the United States as a researcher, inventor, and manager before returning to Russia in 1906. Able to find employment only as manager of a St. Petersburg tram substation, he returned to the United States, where he died in 1923 while working for the Sperry Gyroscope Company.[108]

The lack of financial support directly caused its demise, but even with more funding Lodygin's lamp ultimately would have merely joined the ranks of unsuccessful lightbulbs. The lamp was commercially impractical. It had a short life of several hours (versus the thousand hours of the first Edison bulbs) and operated in small clusters that required their own generating station, a major investment. By contrast, the Edison lamp was the visible part of a complete system, designed from conception to be economically competitive and technically superior to gas lighting.[109] Whereas Lodygin developed a lamp, Edison developed an entire system that demanded minimal investment and attention from the consumer.

The Lodygin lamp would have failed in Russia or Europe. The Jablochkov arc lamp, by contrast, succeeded abroad but failed in Russia. The inventor initially worked in St. Petersburg, but in 1875 he went to Paris, either to flee his creditors or to seek financial support.[110] Certainly, Paris, the international center of electrical engineering in the 1870s, offered a stronger technical base than St. Petersburg.

In an arc light, an electric current passes between two carbon electrodes to generate a bright, intense arc of light. The electrodes must be constantly readjusted as they burn to provide even lighting. The "Jablochkoff candle" solved this problem with electrodes placed side

---

[106] V. L. Chikolev, "Istoriia elektricheskogo osveshcheniia," *Elektrichestvo*, 1880, no. 5: 73.

[107] S. A. Gusev, *Razvitie sovetskoi elektrotekhnicheskoi promyshlennosti* (Moscow: Energiia, 1964), 20.

[108] "Khronika," *Elektrichestvo*, 1923, no. 12: 644–46; M. A. Shatelen, "Lodygin, Jablochkov, Edison, 1847–1947," *Elektrichestvo*, 1947, no. 10: 68–74: "Obituary," *Journal of the AIEE*, May 1923, 553; Bright, *Electric Lamp Industry*, 120–21.

[109] Hughes, *Networks of Power*, 19–20.

[110] Belkind, *Jablochkov*, 84, and "P. N. Jablochkov: Nekrolog," *Elektrotekhnicheskii vestnik*, 1894, no. 4: 121–22.

Jablochkov arc light. Courtesy of the Smithsonian Institution.

by side, separated by an insulating layer of kaolin china, instead of point to point. The sixteen-candlepower lamp offered advantages of simplicity, constant burning, and the ability to run several sets of candles from one generator. Improvements and modifications greatly increased the arc lamp's efficiency and utility for outdoor lighting and large buildings.

Jablochkov's success derived not only from a better idea but also from the better materials and equipment available in the competitive French environment. In 1876–77 alone, four firms introduced carbon electrodes.[111] The Russian benefited from the Gramme generator, significantly lighter, smaller, and cheaper than the previous standard, the Alliance generator. The Gramme was the first generator to achieve commercial success in Europe and Russia.[112] Such improvements enabled Jablochkov to create a lighting system in which all the components and not just the lamp functioned well.[113] In short, Jablochkov integrated the work of others and thought commercially.

Jablochkov's arc light, patented in France in 1876, soon illuminated the streets, public gardens, and factories of Paris, London, and other European cities.[114] The world's first truly commercially successful electric light was easy to use, relatively inexpensive, and reliable. The Jablochkov lamp was not the only Russian arc lamp, but it was the first, the most successful, and the only one backed by a European industrial base, financing, and market.[115] Russian inventors built other lamps, including the Dobrokhotov arc lamps that illuminated Moscow streets in the early 1880s, but the most used arc lamp in Russia—and Europe—was Jablochkov's.[116]

The Russian navy introduced the Jablochkov light to Russia. In 1878, a naval delegation, including five electrical specialists and headed by Gen.-Adm. Konstantin N. Romanov, the tsar's brother, visited the Paris international exhibition. After a demonstration by Jablochkov, Romanov ordered Vladimir P. Verkhovskii, the director

---

[111] Hippolyte Fontaine, *Electric Lighting: A Practical Treatise* (London: E. & F. N. Spon, 1878), 38–50.

[112] James E. Brittain, "The International Diffusion of Electric Power Transmission, 1870–1920," *Journal of Economic History*, March 1974, 108.

[113] D. A. Lachinov, "Poslednye uspekhi v elektricheskom osveshchenii," *ZIRTO*, 1879, no. 2: 77–80.

[114] "The Jablochkoff System of Electric Illumination," *Engineering*, 26 July 1878, 63–65.

[115] Heerding, *Philips' Gloeilampenfabrieken*, 89.

[116] V. Tikhmorov, "Elektricheskaia vystavka: Spiralnaia lampa dlia elektricheskogo osveshcheniia," *Elektrichestvo*, 1882, no. 6: 73–74; N. Sluginov, "Elektricheskaia vystavka: Elektricheskaia lampa A. V. Dobrokhotov-Maikova," *Elektrichestvo*, 1882, no. 7: 88–89; "Deiatelnost obshchestva," *ZIRTO*, 1910, no. 5: 69.

of the Mine Officer Class, to test the inventor's lamp in Russia. The navy convinced Jablochkov to return to Russia and open a factory.[117]

The practicalities of electric light deterred many. The obstacles were high. Owning an arc light required a substantial financial investment for the engine, generator, and other equipment, plus the skilled technical personnel to operate and maintain the equipment. A Jablochkov four-light system cost 1,750 rubles, an insurmountable obstacle for many in 1878.[118] The military had the skilled personnel and the funding, and military orders constituted over two-thirds of the company's initial business. Of the approximately 750 arc lights in Russia in 1881, one-third illuminated military installations and military-related factories and another third illuminated forty ships of the Baltic and Black sea fleets.[119]

Civilian users were either more technically advanced than most firms, such as the Poltava railroad workshop, or, like the Hermitage Gardens, used the lights as a novelty to attract customers. St. Petersburg, site of most of the advanced, Western-related industries, housed nearly 60 percent of the lamps.[120] The largest potential market, city governments, despite illuminating some bridges and squares electrically in St. Petersburg and Moscow, proved reluctant to replace the less costly kerosene and gas lamps for street lighting.[121]

Despite the military orders, Jablochkov's company never really succeeded in Russia. It finally succumbed in 1887 to an inadequate domestic market, manufacturing problems, and competition from foreign firms.[122] Jablochkov himself returned to Paris in 1880, where he continued his research, obtaining thirty-four French patents before he died in 1891.[123]

Jablochkov's major contributions to Russian electrical engineering were twofold. He introduced arc lighting to the country and, more

[117] Belkind, *Jablochkov*, 168–70, 176–77.

[118] *Tovarishchestvo dlia ekspluatatsii elektricheskogo osveshcheniia v Rossii* (St. Petersburg: A. E. Munster, 1878), 12–16; Chikolev, *Lektsii po elektrotekhnike*, 10.

[119] For the survey, "Raznye izvestiia," *Elektrichestvo*, 1882, no. 5: 69; for the fleets, Belkind, *Jablochkov*, 178, 180; Averbukh, *Petrushevskii*, 48.

[120] "Raznye izvestiia," *Elektrichestvo*, 1882, no. 5: 69.

[121] "Elektricheskoe osveshchenie Imperatora Aleksandra II, v S.-Peterburga," *Elektrichestvo*, 1880, no. 2: 24–27; N. I. Falkovskii, *Moskva v istorii tekhniki* (Moscow: Moskovskii rabochii, 1950), 437.

[122] Ie. P. Tveritinov, *Elektricheskoe osveshchenie: Kurs Minnogo ofitserskogo klassa* (St. Petersburg: Morskoe Ministerstvo, 1883), 334; Shatelen, "Lodygin, Jablochkov, Edison," 68–74; Gusev, *Razvitie sovetskoi elektrotekhnicheskoi promyshlennosti*, 11.

[123] M. A. Shatelen, "Pavel Nikolaevich Jablochkov," *Elektrichestvo*, 1926, no. 12: 496–98; "Lists des brevets français pris par P. Jablotchkoff," ibid., 518.

important, in the eyes of fellow entrepreneur Chikolev, "by his energy and labor he cleared the road for other inventors" and brought attention and capital to the Russian electrotechnical industry.[124] Not all this attention was favorable: Jablochkov at times received a curiously hostile reception from the Russian electrotechnical community, possibly because of his international renown.[125]

A third inventor, Dolivo-Dobrovolsky, also spent most of his career outside Russia. Initially, this was involuntary, stemming from his 1878 expulsion from the Riga Polytechnic Institute for political activities. He went to Darmstadt to complete his education and stayed after 1887 to work for the German electrotechnical firm AEG. In 1888, he began research on 3-phase AC transmission. In 1891, he demonstrated long-distance transmission of electricity over the 170 kilometers from Lauffen to Frankfurt, a major technological milestone. Dolivo-Dobrovolsky advanced electrical engineering in Russia from Germany by contributing papers to journals and to the first All-Russian Electrotechnical Congress, donating his library, and supplying equipment to the St. Petersburg Polytechnic Institute. His visits to Russia, however, were short, partly for reasons of health. He declined a position at St. Petersburg Polytechnic Institute to remain with AEG until just before his death in 1919.[126]

The careers of these inventors share several similarities. All spent much of their professional lives in the West, where they achieved greater success than in Russia. Only Lodygin did his major creative work in Russia, work that was ultimately unsuccessful. Mikhail Shatelen explains Lodygin's failure in terms of Russia's poorly developed social-economic base.[127] Shatelen is correct, but the reasons are deeper than he proposes. The West did have the technical base, the financial support, and the market that Russia lacked. But commercial success also demands the full development of all components of a system, including packaging for the consumer. As W. Bernard Carlson and A. J. Millard, biographers of Edison, noted, "success did not

---

[124] Cited in Lachinov, "Poslednye uspekhi v elektricheskom osveshchenii," 89.

[125] If *Elektrichestvo* articles are a guide, Jablochkov had strained relations with his peers. One article stands out. Written by "S. S.," "Novyi element gospodina Jablochkova" (*Elektrichestvo*, 1884, nos. 20–21: 163–64), in addition to the gratuitous title "Mr.," begins with a venomous satire on the inventor. The reader is told that the article is presented solely because of requests; only foreign information is used. Jablochkov's obituary in *Elektrichestvo* (1894, no. 7: 97–99) significantly lacks information about his activities after returning to Russia in 1878.

[126] Oleg N. Veselovskii, *Dolivo-Dobrovolsky, 1862–1919* (Moscow: Izdatelstvo Akademii nauk, 1963); "Nekrolog," *Elektrichestvo*, 1930, no. 5: 258–59.

[127] Shatelen, "Lodygin, Jablochkov, Edison," 68.

come necessarily to the fellow who invented something first. It came to the fellow who could make a new device simple and functional, who could figure out how to manufacture it cheaply and in quantity and then convince people to buy it."[128]

Jablochkov's success can be better understood with Hughes's concept of reverse salients—"obvious weak points, or weak components, in a technology which are in need of further developments."[129] In the West, Jablochkov found the auxiliary technologies and financial support he needed. Like Lodygin, Jablochkov invented a component, not a system. Unlike Lodygin, Jablochkov's French environment provided the other components needed to create a successful lighting system. Unlike Lodygin, Jablochkov utilized French financial and manufacturing support to transfer his laboratory prototypes into commercial products.

The unsuccessful efforts of another Russian, Fedor A. Pirotskii, illuminate the difficulties of the independent Russian inventor and the limits of military interest.[130] An artillery captain, Pirotskii promoted electric power transmission, electric railroads, and electric lighting. In 1874, he proposed a small hydrostation to power a state gunpowder factory. In 1880, the GAU finally offered grudging support of 300 rubles to demonstrate his system of electric transmission, a pittance compared with the tens of thousands of rubles Petrushevskii had spent in his lighting experiments a decade earlier. The project was moderately successful, but it suffered from insulation problems exacerbated by the cold, damp St. Petersburg climate and the erroneous but prevailing assumption that large quantities of electricity demanded a conductor with a large cross section.[131] Like Lodygin, Pirotskii was slightly ahead of the materials and ideas of his time and lacked resources and patrons. His biographer claims that Pirotskii built the world's first electric railroad for the 1880 St. Petersburg electrical exhibition. A Siemens representative reportedly talked to Pirotskii and asked for information about his work, which led to changes in the Siemens electric railroad, first displayed at the 1881 Berlin exposition.[132] In September 1880, Pirotskii did conduct a series of tests at the St. Petersburg horse tram park, which left observers

---

[128] W. Bernard Carlson and A. J. Millard, "Edison as a Manager of Innovation: Lessons for Today," *New Jersey Bell Journal*, winter 1985–86, 27.

[129] Hughes, *Networks of Power*, 22.

[130] For another failure, see N. Popov, "Pamiati A. I. Poleshko," *Elektrichestvo*, 1916, no. 9: 945–47.

[131] TsGIAL f. 506, op. 1, d. 411, 28, 145–50.

[132] B. N. Rzhonsnitskii, *Fedor Apollonovich Pirotskii* (Moscow: Gosenergoizdat, 1969), 45, 55–57.

less than impressed at his "toy."[133] The train suffered from slow speed and costly, unreliable batteries, the same problems that bedeviled engineers in the West. Although Pirotskii thought that he had built an electric railroad before Siemens, his Russian contemporaries ignored his work and credited Siemens for the first electric railroad, as did early Soviet writers.[134]

Pirotskii is interesting, not because he was a military inventor, but because he failed to win acceptance and support from his peers. Part of his failure is not surprising: the GAU served military needs, and Pirotskii's research was not directed to existing needs. Even if his electric railroad had proved practical, what would the GAU have done with it? Siemens, by contrast, was a manufacturing firm creating, shaping, and meeting the needs of customers in the military and civilian spheres. Pirotskii highlights another instance of a technology developed successfully outside and unsuccessfully inside Russia.[135] Russian governmental, financial, and industrial decision makers suffered from a "foreign is better" bias toward technology which handicapped native inventors and firms as Western criteria and activities took precedence over Russian equivalents.[136] In a society where German was the language of the businessman and French the language of the court, this foreign bias is not surprising. This preference for foreign technology and engineers strengthened contacts between Russia and Europe but weakened domestic industrial development.

The activities, ideas, and interests of Russian inventors in the early decades of the electrical industry paced their Western counterparts. In development, diffusion, and application, however, the advantages lay with the more hospitable economic and social environment of the West, with its larger, more advanced technical and financial base. This base, better able to sustain failure and support new ideas and systems, proved the key factor in the rapid Western expansion of electrical applications. The failure of Russian inventors indicates not personal inadequacies but more general societal handicaps.

[133] *Golos*, 17 September 1880, 3; *Russkii invalid*, 16 September 1880, 2.

[134] TsGIAL f. 506, op. 1, d. 411, 70; Iv. Sviatskii, *Istoriia elektrichestva* (St. Petersburg: P. P. Soikin, 1897), 120–21; V. P. Kashchinskii, "Znamenatelnye sobytiia v istorii razvitiia generirovaniia i kanalizatsii elektricheskoi energii za poslednie polveka," *Elektrichestvo*, 1930 Jubilee Issue, 88.

[135] Calling Pirotskii the inventor of the electric tram is misleading; he was one of several inventors around the world working on the same idea at the same time; see John P. McKay, *Tramways and Trolleys: The Rise of Urban Mass Transport in Europe* (Princeton: Princeton University Press, 1976), 35–40.

[136] Rieber, *Merchants and Entrepreneurs*, 102–3.

The Role of Foreign Firms and Investment

FOREIGN INVOLVEMENT was crucial to the industrial development of Russia; envisaging Russia without the large migrations of monies, technologies, ideas, and people from West to East is inconceivable. Although the exact numbers remain a source of contention, foreign investment accounted for significant amounts of government and nongovernment capital formation.[137]

Financing is the underlying sine qua non of commercial technologies. The best equipment in the world is useless without the money to purchase and operate it. High technology did not come cheap, and the Russian financial infrastructure was woefully unsuited to provide the necessary capital.[138] One contributing factor was the tsarist restrictions on the Russian stock exchange, which, by hindering the efficient creation and transfer of capital, increased the country's dependence on foreign capital to finance capital-intensive industries—such as electrification.[139] In the West, financial markets evolved to meet the demand for electric light, power, and traction beginning in the 1880s. The early loans and exchanges of stocks between manufacturers and utilities evolved into banking syndicates, such as the Zurich-based Elektrobank, holding companies, and other mechanisms to transfer equipment to the utilities and profit to the providers.[140] The Russian electrotechnical market did not expand rapidly until the late 1890s and, by then, better-capitalized foreign firms had established Russian subsidiaries that often provided financial support with their technical offerings.

Foreign banks and companies financed the vast bulk of prewar Russian electrification, usually with a Russian bank, especially the Inter-

[137] Arcadius Kahan, "Capital Formation during the Period of Early Industrialization in Russia, 1890–1913," *Cambridge Economic History of Europe*, vol. 7, pt. 2: 273; P. V. Ol, *Foreign Capital in Russia*, trans. Geoffrey Jones and Grigori Gerenstain (New York: Garland Publishing, 1983), 9; Fred V. Cartensen, "Numbers and Reality: A Critique of Foreign Investment Estimates in Tsarist Russia," in Maurice Levy-Leboyer, ed., *La Position internationale de la France* (Paris: L'Ecole des Hautes Etudes en Sciences Sociales, 1977), 275–83.

[138] Trebilcock, *Industrialization of the Continental Powers*, 224–25.

[139] *Potrebitelskie elektricheskie stantsii* (Moscow, 1913), 3; Rieber, *Merchants and Entrepreneurs*, 105.

[140] M. Giterman, "Elektrichestvo i munitsipalitety," *Izvestiia Moskovskoi gorodskoi dumy*, 1914, no. 11: 64; Armstrong and Nelles, *Monopoly's Moment*, 116; Chandler, *Visible Hand*, 310, 426–33; Walther Kirchner, "Siemens and AEG and the Electrification of Russia, 1890–1914," *Jahrbucher für Geschichte Osteuropes* 30 (1982): 408; A. J. Millard, *A Technological Lag: Diffusion of Electrical Technology in England, 1879–1914* (New York: Garland Publishing, 1987), 155–56.

national and Private banks.[141] According to Valentin Diakin, of 139 million rubles invested in utilities by 1914, German monies accounted for nearly half, Belgium-channeled capital for a quarter, Russian funding for about 10 percent and other countries provided the rest.[142] In trams, Belgian firms held 73 percent, Germans 13 percent, and Russians 12 percent of the 94 million ruble investment. German and Belgian firms accounted for 90 percent of the 61 million rubles invested in manufacturing.

No less significant were the flows of foreign technology. Technology transfer took several forms during this half-century, including equipment, such as Parsons turbines, and manufacturing technology, such as factories to produce lightbulbs. Foreign financing and ownership often accompanied these visible forms of technology transfer. The German firm Siemens and Halske dominated Russian manufacturing, in competition with AEG, Brown-Boveri, Westinghouse, Metropolitan Vickers, and other foreign and Russian firms. The strength of German firms lay in their aggressive and thorough marketing. The German businessman in Russia knew Russian, carried brochures and catalogs in Russian, and could arrange long-term credit, a vital consideration.[143] A less visible but important form of technology transfer consisted of "stocks of knowledge," including people, information, and ideas.[144] Foreign companies sent engineers and managers to operate their Russian facilities, train Russians, and sell equipment. Tens of thousands of Russians traveled abroad for technical and scientific training.[145] Trips abroad enabled engineers to meet their Western counterparts and to see and work on the latest technologies.

Electrical engineers proved no exception. Of forty prominent prerevolutionary electrical engineers, two-thirds studied or worked abroad.[146] The tsarist police inadvertently encouraged travel and emigration by restricting and punishing political activities, as in the cases of Dolivo-Dobrovolsy and Achilles de Khotinsky.[147] The Russian government, particularly the military, and technical societies also sent

[141] V. A. Diakin, *Germanskie kapitaly v Rossii* (Leningrad: Nauka, 1971), 41–44, 84–85.

[142] Ibid., 268–69. Diakin excluded an unknown number of municipal operations and domestic concessions, thereby somewhat understating the Russian contribution.

[143] Walther Kirchner, "Russian Tariffs and Foreign Enterprises before 1919: The German Entrepreneur's Perspective," *Journal of European History* 11 (1981): 361–80.

[144] Simon Kuznets, *Toward a Theory of Economic Growth* (New York: Norton, 1968), 62.

[145] Trebilcock, *Industrialization of the Continental Powers*, 268, 290.

[146] Data compiled from *Elektrichestvo* obituaries and *Great Soviet Encyclopedia* articles.

[147] Heerding, *Philips' Gloeilampenfabrieken*, 140, 148.

delegations to Europe for electrotechnical congresses, exhibitions, and factory tours.[148]

Although a major conduit of information about Western electrotechnology, engineers abroad were a small fraction of the Russian electrotechnical community. The majority received information mainly from foreign and Russian periodicals. Graduates of the St. Petersburg Polytechnic Institute in 1913 read thirteen electrical journals. Seven were German, three Russian, and three English or American. Half of these engineers read the German *Elektrotechnische Zeitschrift*; 70 percent read *Elektrichestvo*.[149] Russian electrotechnical periodicals contained numerous translated articles, Russian articles on Western developments, and sections devoted to foreign activities. *Elektrichestvo* began in 1880 with a table of contents in Russian and French. By the late 1880s and 1890s, French articles declined and articles of German and British origin increased. American articles did not reach significant numbers until the 1910s. These changes corresponded to the shift in the frontiers of electrical engineering from Paris to Berlin. A German transfer of knowledge reflected dominance of the Russian electrical market. Even the technical language was German.[150] Fifty-five percent of the St. Petersburg Polytechnic graduates in 1913 knew German; only 28 percent knew English.[151]

The migrations between Europe and Russia included organizational links and ideas. In some areas, Russia–Europe connections proved stronger than intra-Russian ties. A Russian association of utilities did not exist until 1917, but twenty Russian utilities belonged to the Vereinigung Deutscher Elektrizitatewerke, a German association of utilities, in 1914.[152] As the shortages in World War I proved, Russia strongly depended on German electrotechnology and finance. Even the first effort by the VI Section in 1908 to publish statistics on Russian utilities depended on German information.[153] This dependence developed voluntarily; Russian engineers, scientists, and managers

[148] TsGVIA f. 506, op. 1, d. 409, 46–47, 81–85.

[149] M. A. Shatelen, "Iz 'Ankety sredi inzhener-elektrikov' okonchivshikh STP Politekhnicheskii institut Imperatora Petra Velikogo," *Elektrichestvo*, 1914, no. 4: 130.

[150] E. g., Russians used the German *schwachstrom* (weak current) and *starkstrom* (strong current) to distinguish between telecommunications (telegraph and telephone) and the power industry.

[151] Shatelen, "Ankety," 136.

[152] "Khronika," *Elektrichestvo*, 1917, nos. 9–10: 145.

[153] "Statisticheskie svedeniia o tsentralnykh elektricheskikh stantsiiakh v Rossii," *Elektrichestvo*, 1910, no. 1: 1.

saw themselves as part of the larger international community and gravitated toward Germany. The most important foreign society for Russian electrical engineers was the German Verband Deutscher Elektrotechniker. In 1888, 319 of its 1,452 members were non-German.[154] Fifty-four members—17 percent of all foreign members, about half the active membership of the VI Section—were Russians. Non-German societies did not attract similar interest, further proof of the German domination.[155]

Foreign influence permeated every aspect of Russian electrotechnology. The larger European and American bases of production and consumption enabled Western development to create technical, educational, and financial infrastructures that provided commercial advantages abroad in such less developed areas as Russia. Foreign financing permitted Russian electrification to develop as quickly as it did, despite the inadequate Russian credit market. Superior foreign financing provided the means to acquire superior foreign electrotechnology, and Western institutions provided education to Russian engineers. Equally important, the West provided ideas, concepts, and legitimation for Russian electrification proposals that appeared after 1910.

The Russian economic, political, and governmental environment greatly shaped the evolution of technologies in Russia. Electrification was handicapped by a governmental morass that left basic questions unresolved, a time-consuming system for obtaining permission, and government regulations that consistently lagged behind technical developments. National regulations governed the extent and timing of municipal activities, company formation, and the construction and operation of utilities. The structure of rules and reviews slowed the diffusion of new technologies and the creation of utilities. The legal framework hindered the development of indigenous small companies. Larger, better-capitalized foreign firms could endure the time needed to obtain permission and funding more easily than smaller Russian firms. In everyday operations, the process of evaluating proposals for utilities operated sluggishly. For technologies that required new laws, such as hydropower and long-distance transmission, politi-

---

[154] "Raznye izvestiia," *Elektrichestvo*, 1888, nos. 17–18: 176. The destruction of association membership records in World War II makes full knowledge of Russian involvement impossible (VDE personal communication, 26 June 1982).

[155] E. g., Russian membership in the British Institution of Electrical Engineers varied from none to three from 1872 to 1915; data produced by Geoffrey Tweedale for Reader, *History of the Institution of Electrical Engineers*.

cal struggles among ministries slowed or prevented commercial implementation.

The major role played by European firms reflected both Russian weaknesses and foreign advantages in organization, financing, and technology. Capital-intensive electrotechnology fared poorly in risk-adverse, credit-poor, conservative Russia. Compared with the West, Russian electrification advanced quickly in the military sector but more slowly in the civilian sphere, as the poor commercial record of Russian inventions demonstrates. Was the military dominance in the early years of electrification an example of prescience or civilian weakness? Were the Russian army and navy ahead of their time or was the Russian civilian economy behind the times? Similar questions could—and should—be asked of previous and contemporary military research.[156]

The failure of Russian inventors in Russia and the success of some, such as Jablochkov, abroad demonstrates that the receptivity of the environment plays a major role in the invention, development, and diffusion of technologies. This is not a new conclusion, but it bears repeating. Similarly, the major role of the VI Section in the development and transfer of electrotechnology demonstrates the importance of key groups of technical experts. Although few in numbers, the members of the section played vital roles in Russian–Western and intra-Russian technology transfers. In the prewar period, the role of electrical engineers in tsarist policy making was limited to advising. As the importance of electrification increased in the war and postwar periods, so too did the importance of electrical engineers in setting and implementing state policy.

[156] Possibly the best nineteenth-century American example is the four decades of military investment before large-scale manufacturing of truly interchangeable rifles became feasible; see Smith, *Harpers Ferry*, and David A. Hounshell, *From the American System to Mass Production, 1800–1932* (Baltimore: Johns Hopkins University Press, 1984), 15–50.

CHAPTER 3

Electrification,
1886–1914

ELECTRIFICATION GREW SLOWLY in Russia, especially compared with the West. Over a decade passed between the first Russian commercial utility in 1886 and the first large wave of utilities elsewhere in the country. Another decade passed before utilities truly surged into Russian towns and cities. The Russian environment contributed to this slow diffusion, but so did uncertainties about technologies, financing, and organization. Russian decision makers had to respond to the major issues in the electrification of the West: the choice between electricity and other forms of energy; questions about which technology to generate electricity; and debates about the organization, ownership, and financing of utilities and their relation to the government.

A unique national style of electrification emerged from the constraints and opportunities of imperial Russia. The two most obvious technical differences with the West were the lack of hydroelectric power and long-distance transmission, and a utility preference for direct current (DC) over alternating current (AC). Contrary to its image as a backward technology, DC proved a technically feasible and economically sensible choice for Russian utilities. Less apparent but key to the Russian evolution of electrification were a passive national government and its restrictive administrative process, weak local governments, limited financing, and an electrotechnical community often with international ties stronger than domestic links. The country's low level of urbanization influenced the spread of electrification, as did the shape of industrialization.

St. Petersburg and Moscow take center stage in Russian electrification because they spawned the first commercial stations in 1886–87 and remained in the forefront of electrotechnology through the late 1920s. Not until the 1897 St. Petersburg model agreement gave municipalities and potential concessions an administrative and legal *vade mecum* did the spread of utilities truly begin. Technical developments appeared first in St. Petersburg and Moscow and diffused to other cities and towns before spreading to villages and rural areas. By 1914, most cities had a utility, although electric light and power did not touch the daily life of most urban dwellers, let alone the peasantry.

The first visions and proposals of large regional stations to expand the geographic and social base of electrification for economic and political reasons appeared before World War I. These ideas, initially modest, lay the groundwork for future electrification plans and, like later efforts, arose from both foreign influences and indigenous factors, including a growing progressive movement in Russian cities for municipalization. Equally significant, these proposals intertwined with utility interest in regional stations to increase generation and guard against municipalization. Both radical and conventional approaches viewed the regional station as key to Russia's future, but their institutional settings and visions of the future differed greatly.

Early Operations

PRECEDING COMMERCIAL UTILITIES, small privately owned "block" stations provided the first electric arc lighting. Outside the military, the large material, human, and financial investment restricted these lights to a few factories, marketplaces, and public sites despite, as an 1878 brochure suggested, potential application in "large stores, concert halls, restaurants, theaters, hospitals, museums, palaces, monuments, squares, railroad stations, docks, steamships, lighthouses, factories, workshops, night and portable works, naval and military affairs, etc."[1]

The incandescent lamps of Edison and Swan, invented in 1879, opened a much wider market. The incandescent lamp produced a less harsh, less bright light that could be switched on and off at will. By 1882, Russian consumers used Edison, Rene, Siemens, and Swan incandescent lamps to illuminate a variety of public and industrial

[1] *Tovarishchestvo dlia ekspluatatsii elektricheskogo osveshcheniia v Rossii* (St. Petersburg: A. E. Munster, 1878), cover.

structures, including St. Isaac's Cathedral, a regimental meeting hall, the Petrovsk theater, and the Novorossiisk oil terminal.[2]

Incandescent and arc lights faced major challenges of high cost and competition from other lighting sources. The main competitors were kerosene and stearin candles. Although technically inferior to electricity, kerosene cost far less because of abundant, inexpensive supplies of Caucasian oil. Kerosene also had the advantages of low installation costs, bureaucratic acceptance, and few technical demands on users. An electric street lamp cost seven to twenty times more to install than its kerosene equivalent.[3] In an example of regulations serving as a ceiling and not a floor, standards for lighting government buildings, developed for kerosene, hindered efforts to electrify municipal and state facilities.[4] Kerosene lighting also did not demand the expensive supporting technical infrastructure and skills required for gas or electricity. Electric lamps were fixed in place and dependent on an outside source of power; kerosene lamps were filled by the owner and placed where desired.

In St. Petersburg and Moscow, electric lighting competed against gas lighting, a technology developed in the 1830s.[5] Advocates of electricity proclaimed that their lighting "burns extremely evenly, gives less heat, does not spoil the air, and does not hiss."[6] Furthermore, electric lighting preserved building interiors because it did not produce damp and sulfurous fumes, carbon dioxide, and soot.[7] For factories where a spark or open flame could ignite a fire, electric light was safer than gas or kerosene. Although it was the major competitor of electric lighting in the West, gas illuminated fewer Russian cities than did electricity.[8] It nonetheless remained a competitor because of the spread of gas motors, increases in the cost of kerosene, and substantial improvements in gas burners in response to the challenge of elec-

[2] Respectively, in *Elektrichestvo*, A. Lazerev, "Elektricheskoe osveshchenie Isaakievskogo sobora," 1883, nos. 10–11: 125–26; F. Kresten, "Elektricheskoe osveshchenie lampami s nakalivaniam," 1884, nos. 17–18: 141–46; "Raznye izvestiia," 1882, no. 2: 29; nos. 10–11: 65.

[3] "Naruzhnoe osveshchenie," *Entsiklopediia mestnogo upravleniia i khoziaistva* (Moscow-Leningrad, 1927), 546, 550.

[4] "Doklad gorodskoi upravy," *Izvestiia S.-Peterburgskoi gorodskoi dumy*, 1908, no. 24: 2242–43.

[5] N. Ivanov, "Sto let gazovogo dela," *Izvestiia Moskovskoi gorodskoi dumy*, 1909, no. 5: 38–46.

[6] "Raznye izvestiia," *Elektrichestvo*, 1882, nos. 3–4: 52.

[7] A. Lazerev, "Elektricheskoe osveshchenie Isaakievskogo sobora," *Elektrichestvo*, 1883, nos. 10–11: 125–26.

[8] Thirty-three cities in 1904 and 104 in 1910; "Naruzhnoe osveshchenie," 551.

tricity.[9] Such increased innovation is a standard response by an old technology under assault from a new technology.[10]

Kerosene lamps illuminated the streets of 80 percent of Russia's cost-conscious municipalities, compared with 14 percent for electricity and 10 percent for gas in 1910. In St. Petersburg, 47 percent of the 18,000 lamps in 1914 were gas, 37 percent kerosene, and 16 percent electric.[11] Russian cities differed from Western cities not in the low percentage of electrified lights—St. Petersburg had more arc lamps absolutely and by percentage than Paris or Berlin—but in the use of kerosene lamps. Even in 1910, most town governments considered electric lighting as the street light for the future.[12]

In 1880, electricity suffered from high cost and extensive accompanying technical equipment compared with kerosene; electricity's advantages were a better quality of light, safety, and hygiene. Three decades later, kerosene was still the main competitor, cost the major objection to electric lighting, and ignorance still widespread.[13]

Commercial Electric Power, 1883–1914

THE LATER DIFFUSION of utilities and their small size distinguished Russia from the West. Compared with the United States, where electric stations spread like wildfire—815 municipal stations in 1902 and 1,562 a decade later—tsarist Russia moved much more slowly.[14] Although the number of prewar utilities varies from 100 to 500 depending on definition, a realistic estimate for 1914 is 200–250 utilities, of which 50–70 percent were concessions.[15] More important, only 12 (5

[9] E. g., a new burner introduced in 1909 gave a fourfold increase in light at one-third the cost of the old burner; "Moskva," *Gorodskoe delo*, 1911, no. 2: 163. See also Ivanov, "Sto let," 40.

[10] Nathan Rosenberg, *Perspectives on Technology* (New York: Cambridge University Press, 1976), 202–6.

[11] "Naruzhnoe osveshchenie," 551; "Doklad o sposobakh uluchsheniia finansov goroda S.Peterburga," *Izvestiia S.-Peterburgskoi gorodskoi dumy*, 1914, no. 1: 114.

[12] O. G. Flekkel, "VI Vserossiiskii elektrotekhnicheskii sezd," *Gorodskoe delo*, 1911, no. 5: 455.

[13] A 1907 congress on lighting and heating featured forty-three kerosene and twenty-six electrical exhibits; see V. I. Kovalevskii, "Osnovnye zadachi iskusstvennogo osveshcheniia," *ZIRTO*, 1908, no. 2: 88. See also I. Shirman, "Ustroistvo elektricheskikh stantsii gorodskimi upravleniiami," *Gorodskoe delo*, 1909, no. 15: 748; "Po Rossii," *Ekonomicheskaia zhizn*, 1 January 1919, 6.

[14] "Central Stations in the United States," *Electrical World*, 14 March 1914, 586.

[15] "Zapiska VI (elektr.) otdela IRTO po voprosu ob oblozhenii aktsiz elektricheskoi energii, idushchei dlia tseli osveshcheniia," *Elektrichestvo*, 1915, no. 1: 21–29; "Spisok

percent) Russian stations had a capacity over 5 megawatts (MW) in 1914, compared with 162 (10 percent) in the United States and 103 (3 percent) of Germany's 4,040 stations.[16] This vast internal disparity in station size is confirmed by Soviet statistics, which list 230 stations in 1913 with a capacity of 328 MW, of which 221 urban stations had only 151 MW compared with 177 MW in the nine stations in St. Petersburg, Moscow, and Baku.[17] As Table 3.1 demonstrates, Russia had fewer stations providing less electricity to fewer customers than did the West.

Russian urban areas can be divided into three tiers of electrification: cities of more than 250,000 that had at least one utility; towns and cities of 50,000–250,000 that probably had a utility; and towns below 50,000, which had a 5–10 percent likelihood of possessing a utility (see Table 3.2). The tiers reflect significant differences in capacity, but also in load factor, diffusion of utilities, tariffs, ownership, technology, and, after 1917, governmental regulation and administration. In general, the larger the city, the more likely it was to have an industrial load, lower tariffs, concessionary ownership, AC rather than DC, turbines rather than less advanced compound engines, and earlier electrification. The three tiers do not quite parallel these population categories because the gap in utility capacity between St. Petersburg, Moscow, and Baku and the other cities over 250,000—Kharkov, Kiev, Lodz, Odessa, Riga, and Warsaw—was so great that the latter remain in the second tier. In 1913, the nine first-tier utilities of St. Petersburg, Moscow, and Baku contained 177 MW, 54 percent of the 327 MW of all Russian utilities.[18] In contrast to this station average of 20 MW, the six next largest cities contained only 52 MW in nine utilities for an average of 6 MW, or the approximate total of fifty-one other utilities in the second tier, whose stations ranged from a few hundred kilowatts to a few megawatts. Of these sixty stations about which we have detailed information, twenty (33 percent) had less than 500 kW

---

elektricheskikh stantsii, o kotorykh v VI-m otdele imeiutsia svedeniia," ibid., 29–30; and "Biulleten," *Elektrichestvo*, 1927, no. 1: 43. The annual statistical tables in *Elektrichestvo* provide the most extensive data, but they do not include all utilities or full data on every utility. Consequently, the number of stations cited may vary. The annual tables capture about half of all utilities but four-fifths of all capacity, implying that the missing towns and cities were probably smaller than those in the *Elektrichestvo* surveys.

[16] "Statisticheskie svedeniia o tsentralnykh elektricheskikh stantsiiakh v Rossii za 1914 god," *Elektrichestvo*, 1917, nos. 4–6: 56–73; "German Central-Station Statistics," *Electrical World*, 10 January 1914, 105–6.

[17] "Biulleten," *Elektrichestvo*, 1927, no. 1: 43.

[18] Ibid.

Table 3.1. Electric generation and usage, national comparisons

Country	Stations	Installed MW	kWh/person
Russia (1913)	220	300	16
Germany (1913)	4,040	2,100	320
Britain (1912)	568	1,240	—
Sweden (n.d.)[a]	440	7,191	1,300
U.S. (1912)	5,221	5,135	500

Source: L. Dreier, *Zadachi i razvitie elektrotekhniki* (Moscow, 1919), 8.

[a] Sweden's high per capita consumption resulted from the extensive exploitation of hydroelectric power for industry.

of capacity, sixteen (27 percent) had between 500 kW and 1 MW of capacity, nineteen (32 percent) had 1–5 MW, and only five (8 percent) had more than 5 MW. Third-tier utilities ranged from 40 to 600 kW, with a 200-kW average.[19]

Electricity spread widely during these years, but the vast majority of Russia remained unelectrified. What was not done should not be forgotten.

The First Tier: St. Petersburg, Moscow, and Baku

ST. PETERSBURG AND MOSCOW, the country's two largest cities, and Baku, the center of the oil industry, had the only utilities to reach Hughes's stage of competition and consolidation. Each city evolved

Table 3.2. Russian utilities in 1910 by city size

Population (thousands)	Stations			Cities without electricity[a]	Total
	Private	Public	Sum		
<50	6	13	19	367 (95)[a]	386
50–250	22	14	36	20 (36)	56
250–2,000	7	2	9	0 —	9
	35	29	64	387 (86)	451

Source: V. V. Dmitriev, *Doklad II-mu sezdu lits okonchivskikh Elektrotekhnicheskii institut Imperatora Aleksandra III-go* (St. Petersburg, 1910), 4.

[a] Percentage of total in parentheses.

[19] "Statisticheskie svedeniia . . . za 1914 god," *Elektrichestvo*, 1917, nos. 4–6: 56–100.

differently, but together their experiences guided the development of other Russian cities and towns.

The first attempts to provide public electric lighting preceded successful implementation by several years. In 1880, GAU Capt. Pirotskii proposed to illuminate St. Petersburg. Asked by the city to examine Pirotskii's proposal, the VI Section, in one of its earliest acts, formed a commission that included Lodygin and GAU Gen. N. M. Alekseev.[20] Pirotskii's proposal never reappeared. An 1884 proposal suffered a similar fate.[21] In 1885, the St. Petersburg *gradonalchik* (appointed mayor) formed a commission to study electric lighting, again with VI Section participation.[22]

Independent of these efforts, Chikolev's company, Elektrotekhnik, had obtained permission in 1880 to light Nevskii Prospekt, the city's main boulevard. Elektrotekhnik constructed a network of arc lamps but ran out of funding in 1883.[23] The Russian branch of Siemens and Halske bought Chikolev out and completed the project. In late December 1883, thirty-two lamps with 1,200 candlepower provided a brightness such that "in every point of Nevskii [Prospekt] it was possible to read easily."[24]

Siemens and Halske had entered Russia in 1853 to construct telegraph lines for the state. The St. Petersburg–based firm gradually expanded into electrical manufacturing, copper mining, and other industries.[25] Nevskii Prospekt marked its debute into electric lighting. In August 1885, the firm petitioned the Ministry of Finance to form a separate company for electric lighting. Eleven months later, the ministry approved the establishment of the Obshchestvo Elektricheskogo Osveshcheniia (Company for Electric Lighting), commonly known as the 1886 Company, with a basic capital of 1 million rubles.[26] German stockholders, including the company's president, Karl Siemens, predominated.

The city duma voted in December 1886 to sign a lighting contract

[20] "Zhurnal chlenov VI otdela IRTO, 10-go sentiabria 1880 goda," *Elektrichestvo*, 1880, no. 7: 103; Viktor V. Danilevsky, *Russkaia tekhnika* (Leningrad: Gospolitizdat, 1948), 394.

[21] "Otchet o deiatelnosti IRTO za 1884 g.," *ZIRTO*, 1885, no. 1: 113.

[22] "Programma predpolagaemykh zaniatii VI otdela IRTO v 1886 god," *ZIRTO*, 1886, no. 1: 201.

[23] "Raznye izvestiia," *Elektrichestvo*, 1882, no. 7: 96; M. O. Kamenetskii, *Pervye russkie elektrostantsii* (Leningrad: Gosenergoizdat, 1951), 28; also, V. A. Diakin, *Germanskie kapitaly v Rossii* (Leningrad: Nauka, 1971), 21.

[24] "Raznye izvestiia," *Elektrichestvo*, 1883, nos. 21–22: 239; nos. 23–24: 255.

[25] Walther Kirchner, "The Industrialization of Russia and the Siemens Firm, 1853–1890," *Jahrbucher für Geschichte Osteuropas* 22 (1974): 321–57.

[26] TsGIAL f. 20, op. 4, ed. kh. 3594, 1, 73.

with the company.[27] The twelve-year contract, far shorter than the twenty- and forty-year contracts common in the West, offered little time for the utility to recoup its investment. Even before the contract, the 1886 Company operated two stations, one on Kazan Square for private and business subscribers, the second on a wooden barge on the Moika River to light Nevskii Prospekt and the surrounding streets. The standard Siemens two-wire 120–130-volt DC system set the parameters for the next decade. George Cutter, Elihu Thomson's European agent, called the technical arrangement "the best I have seen for a long time."[28] In 1889, an iron barge held a station at the Fontanka Bridge, and a fourth station provided additional street lighting.

Although the immediate impetus was Russian Orthodox opposition to a station near a church, locating a power plant on a barge made excellent sense because the transmission technology of the late 1880s limited range to several hundred meters.[29] The river location kept the station close to customers without the cost of land. The location also simplified the water supply for steam and cooling. The increasing size of stations and improved transmission distances of several kilometers made barges impractical and unnecessary by the late 1890s, although riverside locations remained important for access to water and fuel.

The capital's four gas companies provided the only real competition until the late 1890s.[30] They employed several contradictory strategies—opposition, cooperation, and building their own electric stations[31]—but maintained their position only by industrial sales.[32] Nonetheless, only one gas firm survived the prewar municipal buyouts.[33]

In Moscow, the duma initially moved quickly to introduce electric lighting, but unsatisfactory results prompted a more cautious ap-

[27] *Birzhevye vedomosti*, 30 and 31 December 1886, 2.

[28] George Cutter to Elihu Thomson, 26 March 1886, in Harold J. Abrahams and Marion B. Savin, eds., *Selections from the Scientific Correspondence of Elihu Thomson* (Cambridge: MIT Press, 1971), 210.

[29] Andrei M. Ivanov, *Nevskie ogni: Iz istorii ulichnogo osveshcheniia Peterburga-Leningrada* (Leningrad: Lenizdat, 1969), 23.

[30] The two most important were the French Company for the Gas Lighting of St. Petersburg, established in 1835, and the Company for Street Lighting, established in 1858.

[31] As did gas companies elsewhere: in 1888, fifty-eight American gas companies had electric stations; see Harold C. Passer, *The Electrical Manufacturers, 1875–1900* (Cambridge: Harvard University Press, 1953), 37.

[32] See the annual statements in *Vestnik finansov, promyshlennosti i torgovli* (e.g., 1886: 621–23); see also Kamenetskii, *Pervye russkie elektrostantsii*, 38–46.

[33] "O munitsipalizatsii gazovogo predpriiatiia v Peterburge," *Gorodskoe delo*, 1914, no. 8: 478–85.

proach. In 1880, the city signed a contract with Jablochkov's firm to illuminate the Church of Christ the Saviour and other places, but it annulled the agreement after the company failed to complete the work, apparently because the inventor went to Paris. A second contractor, Kleiber, failed to install Chikolev arc lights because the duma did not advance him funding. Finally, Chikolev's company completed the task early in 1883, but the *uprava* (city executive board) entrusted two officers from the navy's Mine Officer Class to modify and operate the system.[34] Another firm, the Moscow Company for Electric Lighting, was formed in 1883, but apparently it failed to raise the half-million rubles specified in its statutes.[35]

Unlike St. Petersburg, Moscow faced a major technical choice. In 1887–88, the city government nearly signed a lighting contract with the Austrian firm Ganz instead of the 1886 Company. Before proposing in March 1887 to use the Moscow River for hydroelectric power, Ganz tried to combine forces with the DC-oriented 1886 Company, but the latter declined.[36] Ganz claimed that the virtues of its system included low cost, an effective radius of ten kilometers with a 3,000-volt AC transmission line (an order of magnitude greater than DC systems), and thirty-four operating installations, including one in Odessa.[37] The ensuing debate was the first Russian "battle of the systems," a battle decided by fear and conservatism.[38]

In the 1880s and 1890s, advocates of AC and DC fought in the West on many, often emotional, levels as competing companies promoted their systems.[39] Development of AC systems lagged behind that of DC systems, which electrified densely populated urban districts, where DC's low transmission efficiency did not matter. Early 1-phase AC stations electrified more suburban, less dense areas but could power only limited industrial applications. Not until the development of 3-phase AC motors and demonstrations of long-distance transmission

[34] "Raznye izvestiia," *Elektrichestvo*, 1883, no. 9: 117; N. I. Falkovskii, *Moskva v istorii tekhniki* (Moscow: Moskovskii rabochii, 1950), 437; Kamenetskii, *Pervye russkie elektrostantsii*, 31; L. P. Kopylova, ed., *Nesushnyi svet* (Moscow: Profizdat, 1969), 7–8.

[35] *Polnoe sobranie zakonov*, 1883, 1566. I am indebted to Thomas C. Owen for bringing the company to my attention.

[36] Kamenetskii, *Pervye russkie elektrostantsii*, 74.

[37] TsGIAL f. 90, op. 1, ed. kh. 455, December 1887 brochure, 57–71; G. D. Polizo, "Pervaia v Rossii tsentralnaia elektrostantsiia peremennogo toka," *Elektrichestvo*, 1967, no. 12: 79–80.

[38] Unless specified, the sources for this debate are 23 and 26 June 1888 articles in *Moskovskie vedomosti*, in TsGIAL f. 90, op. 1, ed. kh. 455, 56–57.

[39] Thomas P. Hughes, *Networks of Power* (Baltimore: Johns Hopkins University Press, 1983), 106–39.

(by Dolivo-Dobrovolsky from Lauffen to Frankfurt) and hydropower (Niagara Falls) in the mid-1890s did AC's technical and economic advantages become accepted for large utilities. In the mid-1880s, the debate was very much alive.

As in the West, DC advocates in the Moscow debate emphasized fear—or safety.[40] Meingardt, a city engineer, stated that DC and low voltage were significantly safer than AC and high voltage. He concluded that the only purpose of the high-voltage AC line was to save money and increase company profits, charges he did not alone make.[41] In June 1888, the city uprava decided that the Ganz system was too risky for Moscow. The decision was undoubtedly influenced by the 1886 Company's declaration in April to restrict its operations to 150 volts because high-voltage AC "presented an enormous danger to the lives of people." The company asked the VI Section to support its claims, but the section, citing inadequate information and discussion, cautiously agreed only to study the matter.[42] The 1886 Company opposition to AC forestalled a competitor, but the company also feared that the accidents and deaths from AC, publicized in the West as part of the AC–DC commercial struggle, would discredit the industry as a whole. Asking people to differentiate between good and bad electricity was overly optimistic when most had never heard of electricity.

Ganz presented the only serious competition to an 1886 Company monopoly in Moscow. In April 1887, just after the Ganz proposal, the 1886 Company obtained a concession to lay wires underground and provide electric lighting. The expensive 65 kopecks per kWh tariff restricted its market to business and the upper strata of society. Public users, such as trading arcades and commercial buildings, provided the majority of the company's clientele: of the estimated 500 arc lights and 40,000 incandescent lights in St. Petersburg and Moscow in 1887, individuals owned only 3,000.[43]

The 1886 Company bought out small private stations in Moscow and then served their users from two new, larger stations, including the Bolshoi Dmitrov power station, sited at a former monastery.[44] Starting from an initial capacity of 150 kW in 1887, the 1886 Company doubled its capacity roughly every two years to reach 1.5 MW by

[40] Terry S. Reynolds and Theodore Bernstein, "The Damnable Alternating Current," *Proceedings of the IEEE* 64 (September 1976): 1340.

[41] TsGIAL f. 90, op. 1, ed. kh. 455, 56a.

[42] Ibid., 32, 40–41, 56a; Kamenetskii, *Pervye russkie elektrostantsii*, 74.

[43] TsGIAL f. 20, op. 4, ed. kh. 3594, 1, 77.

[44] V. D. Kirpichnikov, "Razvitie Moskovskoi tsentralnoi elektricheskoi stantsii O-va 1886 goda," *Elektrichestvo*, 1914, no. 3: 81; Falkovskii, *Moskva v istorii tekhniki*, 437.

Old machine hall of the 1886 Company's Moscow station. Courtesy of the Soviet Polytechnic Museum.

1894.[45] Gross profits averaged 25 percent during the company's first decade, but expanding and maintaining a capital-intensive network technology demanded reinvestment and new investment.[46] Consequently, the company issued only five dividends, averaging 4.5 percent, in its first decade, although the 1890s proved more profitable than the 1880s.[47] Operations in St. Petersburg produced revenues 50–100 percent greater than in Moscow due to greater output and higher tariffs.

The two cities took very different routes to renewing the 1886 Com-

[45] For 1887, see "Khronika," *Elektrotekhnicheskoe delo*, 1914, no. 4: 19; for 1890 and 1894, see Kamenetskii, *Pervye russkie elektrostantsii*, 43–48.

[46] TsGIAL f. 20, op. 4, ed. kh. 3594, 75, 154–56.

[47] E.g., in 1893–94 an income of 836,510 rubles yielded a profit of 239,496 rubles, of which 130,318 rubles (55 percent) went for equipment. The company kept 11,000 rubles as reserve capital and distributed 90,000 rubles as dividends for a yield of 11 percent; *Vestnik finansov*, 1894: 1042–43.

pany concessions in the mid-1890s. The company remained the major firm and radically revamped its operations in both cities, switching from a low-voltage DC system to 3-phase AC with 2,100-volt main transmission lines and transformers to reduce the current to 120 volts for actual use.[48] Thus, a decade after opposing high-voltage AC—a decade in which Siemens's German rival, AEG, successfully and safely deployed long-distance transmission technology—the 1886 Company switched to AC after receiving the concessionary agreements that encouraged long-term investment and conceding the inability of DC systems to expand geographic coverage efficiently.[49] In Moscow, the company retained its monopoly. St. Petersburg promoted competition from other utilities. Both cities attempted to mandate good service at low cost for all consumers. The different approaches—monopoly and competition—illustrate the problems of controlling a capital-intensive network technology. Theoretically, a single concession best utilizes resources and prevents duplication at the disadvantage of user dependence on one supplier. High prices are an obvious point of dissatisfaction, but the more serious problem, particularly from an industrial viewpoint, is inadequate service. What happens if the concessionaire does not meet demand? Building a power plant solves that problem, but only large firms can afford the investment. Two paths to reap the advantages of monopoly while providing some consumer protection are municipal ownership and governmental oversight of private firms.

In September 1895, the Moscow city government signed a fifty-year concession with the 1886 Company, obligating the company to supply electricity citywide.[50] The contract fixed voltage at 122–124 volts, a measure of standardization from the previously delivered 105, 125, and 155 volts. The contract set rates at 50 kopecks per kWh for lighting and 35 kopecks per kWh for technical and other uses, with sliding discounts for large users. If dividends exceeded 8 percent, the company was to share the additional profits with consumers.[51] At the end of the concession, the physical plant would transfer to the city without compensation.

St. Petersburg decided that the advantages of competition out-

---

[48] *Tsentralnye stantsii Obshchestva elektricheskogo osveshcheniia 1886 g. v Moskve i S.Peterburge* (Moscow: 1886 Co., 1901), 3–4.

[49] Kamenetskii, *Pervye russkie elektrostantsii*, 80–84.

[50] See TsGIAMO f. 722, op. 1., d. 392, 11–12 for the agreement.

[51] In 1911, a dividend over 8 percent brought a tariff cut of 1 kopeck per kWh, or about 5 percent of the 22 kopeck rate; "Khronika i melkie zametki," *Elektrotekhnicheskoe delo*, 1911, no. 3: 22.

weighed the potential of too many inefficient utilities. The 1886 Company supplied only part of the city, so competition promised faster introduction of electric light and power at lower rates. The threat of competition was expected to provide a lever for a more favorable contract and allow the city to judge the performance of each utility comparatively. But experience proved otherwise. From 1890 to 1896, the city signed agreements with six other firms to provide electricity: the Russian Company for Electric Energy and the Gue and Shmattser Company (both bought out by the Belgian Company for Electric Lighting in 1897–99), the German-owned Helios, two gas companies, the firm of Nikolai V. Smirnov, and an insurance firm.[52] In 1895, the city duma decided not to renew the single concession to the 1886 Company and directed the uprava to work out a model agreement with the company and other potential concessionaires. In February 1897, the duma approved the agreement, which gave the city better terms than originally proposed by the 1886 Company.[53] The concession lasted for forty years, after which the utility would belong to the city. The city had opportunities to buy the concession after twenty and thirty years, with compensation based on the profits for the previous five years.[54]

The municipal government renegotiated its concessions on the basis of the model agreement in 1897–98. Buyouts and the 1899–1902 recession eliminated three concessions. By 1902, four utilities provided 16 MW.[55] The smallest utility supplied only 800 kW and was owned by Nikolai V. Smirnov, a retired army colonel who received a concession in 1894 after the 1886 Company failed to fulfill its 1889 pledge to electrify Vasilevskoe Island. His concession required four years to obtain 800,000 rubles, demonstrating the conservative nature and limited funds of the domestic capital market. The firm never became a major utility, and Smirnov soon left to operate a lighting concession in Rostov-on-Don.[56] By 1899, the 1886 Company remained the largest utility but faced potential competition from Helios and the Belgium Company for Electric Lighting.

[52] *Elektricheskie stantsii v S.-Peterburge* (St. Petersburg: Spb. Gradonachalstvo, 1900), 9; "O finansovykh rezultatakh ekspluatatsii na gorodskikh gazovykh zavodov za vremia s 10 oktiabria po 1 dekabria 1908 goda," *Izvestiia S.-Peterburgskoi gorodskoi dumy*, January 1909, 681.
[53] "Doklad gorodskoi upravy," *Izvestiia S.-Peterburgskoi gorodskoi dumy*, March 1914, 2480; Diakin, *Germanskie kapitaly*, 37–38, 108.
[54] *Elektricheskie stantsii v S.-Peterburge*, 11–12.
[55] T. F. Makarev, "Razvitie oborudovaniia tsentralnykh elektricheskikh stantsii v Peterburge," *Elektrichestvo*, 1912, no. 6: 181.
[56] TsGIAL f. 20, op. 4, d. 4058; Kamenetskii, *Pervye russkie elektrostantsii*, 66.

St. Petersburg had hoped that these three utilities would compete citywide. Instead, they formed a syndicate to divide the city into geographic spheres of interest, frustrating the city's intent.[57] From the utilities' viewpoint, this arrangement avoided a potentially ruinous direct competition while providing opportunities for growth and profit. Instead of a choice among competing firms, the consumer was enslaved to the firm covering his area. The city wanted competition by service; it received monopoly by geography. Friction also arose from utility interpretations of the agreement which minimized municipal oversight and rebates.[58] Even worse, the three utilities and the tram station operated on different standards, turning the city into a pastiche of incompatible stations similar to that enveloping London in the 1890s and slowing its growth.[59] In a major oversight, the model agreement had not specified frequency, current, or voltage, preventing an interchangeable and uniform electricity.[60] The utilities and the municipal tram all operated on AC, but voltage and frequency differed, so users served by one utility could not directly use electricity from another. Only the 1886 Company and the tram station generated the 3-phase AC necessary for industrial motors. The result was a city geographically divided into technically incompatible areas and lacking a unified city grid.[61]

Concessionaires operated only six of St. Petersburg's 284 electric stations in 1899. Of the remaining 278, only fourteen stations produced AC power. Their larger size, an average of 110 kW versus an overall average of 70 kW, indicates industrial and, in at least one case, military use.[62] Private stations were spatially distributed by size and function: larger stations were located on the outskirts of the city to serve industries; smaller stations were situated closer in and used for lighting.[63] In 1898, private stations powered 75 percent of the incan-

---

[57] "Doklad gorodskoi upravy," *Izvestiia S.-Peterburgskoi gorodskoi dumy*, March 1914, 2491; V. I. Bobykin, *Formirovanie finansovogo kapitala v Rossii, konets XIX v.-1908 g.* (Moscow: Nauka, 1984), 248.

[58] "Doklad gorodskoi upravy," *Izvestiia S.-Peterburgskoi gorodskoi dumy*, March 1914, 2475, 2490–93, 2506–8.

[59] Leslie Hannah, *Electricity before Nationalization* (Baltimore: Johns Hopkins University Press, 1979), 47–48. For a map, see opposite page 108 in John B. Verity, *Electricity up to Date for Light, Power, and Traction* (London: Frederick Warne, 1891).

[60] "Doklad gorodskoi upravy," *Izvestiia S.-Peterburgskoi gorodskoi dumy*, March 1914, 2522–23.

[61] P. Gurevich, "Angliiskaia elektrotekhnicheskaia promyshlennost," *Elektrichestvo*, 1915, no. 15: 346.

[62] *Elektricheskie stantsii v S.-Peterburge*, 5, 9.

[63] E. g., the twenty-four stations in the Moscow section of the inner city averaged 33 kW and the fourteen stations in the outlying Shlisselburg section averaged 190 kW.

descent and 60 percent of the arc lamps.[64] As central stations expanded geographically, they supplanted the smaller stations by offering lower costs, but did not eliminate private industrial stations.[65]

Major growth began after 1904 in St. Petersburg and Moscow with a big boost from industrial use of utility-provided electricity. Factories converted to electricity from steam power or switched from their own stations to a utility. The industrial load of the 1886 Company took a decade to double from 17 percent in 1900 to 34 percent in 1910, but it needed only four more years to double again to 66 percent in 1914.[66]

Consumption of electric energy expanded rapidly in St. Petersburg from 100 MkWh in 1911 to 148 MkWh in 1914. Industrial demand accounted for most of the growth, but the market for lighting also increased as users grew from 56,000 in 1911 to 83,000 in 1914.[67] Taken collectively, the St. Petersburg utilities generated more electricity than in Moscow, albeit with a lesser overall industrial load because only the 1886 Company provided electric energy in the form most suitable for engines. The 1886 Company moved from generating a plurality of electric power to a majority after 1913.

In 1906, the 1886 Company obtained the concession for the industrial city of Lodz, the "Manchester of Poland," from its parent Siemens and Halske. The manufacturer had obtained the concession in 1900 during a recession but waited until better economic times to develop it.[68] The 1886 Company thus controlled three of the empire's four largest utilities in rapidly growing cities, growth that produced dividends of 6–10 percent but also necessitated constant reinvestment, new stock offerings, and new equipment. As Table 3.3 indicates, the 1886 Company expanded greatly during the post-1907 economic boom.

In 1897, the Moscow section of the 1886 Company became the first Russian utility to use 3-phase AC. The section was directed by Marxist Robert E. Klasson, who had worked under Dolivo-Dobrovolsky and constructed the first large Russian 3-phase AC station for the

---

[64] "Khronika," *Elektrotekhnik*, 1897–98, no. 19: 1142–43.

[65] In the two years after a central station opened in 1896, eight of the nine private stations in the area of the Ligov and Kriukov canals closed; *Elektricheskie stantsii v S.-Peterburge*, 6.

[66] TsGIAL f. 115, op. 1, d. 30, 12, 22; *Elektricheskie stantsii v S.-Peterburge*, 6; "Statisticheskie svedeniia o tsentralnykh elektricheskikh stantsiiakh v Rossii," *Elektrichestvo*, 1912, no. 1: 2, 11. For 1914, see "Statisticheskie svedeniia . . . za 1914 god," *Elektrichestvo*, 1917, nos. 4–6: 98, 100.

[67] "Statisticheskie svedeniia . . . v Rossii," *Elektrichestvo*, 1913, no. 10: 302–10; 1917, nos. 4–6: 88–102.

[68] Diakin, *Germanskie kapitaly*, 102–3.

Table 3.3. Total 1886 Company growth, 1904–1913

	1904	1908[a]	1913[a]
Users	9,401	19,659 (210)	78,035 (400)
MkWh	15.6	35.4 (225)	168.3 (475)
Cables (km)	508	795 (145)	2,645 (330)
Capital (million rubles)	18	18 (0)	40 (220)
Dividend (%)	5.8	8 (15)	8.7 (10)

Source: "Po russkim gorodam," *Elektrichestvo*, 1915, no. 1: 32.
[a] Percentage increase from previous year in parentheses.

army in 1895.[69] The Georgievsk station served as the section's keystone for two decades, changing from vertical compound engines in 1897 to the first turbines in 1902 and more powerful turbines in 1907. The station was to be exceptionally important in the development of Russian electrotechnology through its leadership in training Russian engineers and technicians and in developing and transferring new equipment and methods.[70]

Monopoly guaranteed smooth sailing for neither the utility, the user, nor the municipality. During the early years of the new contract, the Moscow government and the 1886 Company repeatedly clashed over interpretations. Resolution demanded clarification, negotiation, and patience.[71] In 1908, the uprava sought to renegotiate the agreement to increase the city's share of the company's income, improve service, and develop the aesthetics of the company's operations. These discussions evolved into larger debates about electrification which involved two potential competitors, Pavel N. Rattner's St. Petersburg Company for Electric Construction and a Belgian newcomer, I. Fain's Company for Electric Central Stations. By 1911, the duma had rejected the tenets of a monopoly by concession and opened Moscow to competition. Its proposed contract gave the municipality more authority and income while demanding a larger utility commitment.[72] The 1886 Company did not sign this new contract, but it realized that the end of its monopoly loomed.

[69] S. A. Gusev, "Pervaia promyshlennaia ustanovka trekhfaznogo toka v Rossii," *Trudy po istorii tekhniki* 6 (1953): 74–84; M. O. Kamenetskii, *Robert Eduardovich Klasson* (Moscow: Gosenergoizdat, 1963), 22, 25–32, 50–54.

[70] TsGANKh f. 9508, op. 1, ed. kh. 14, 4.

[71] TsGIAMO f. 722, op. 1, d. 392 covers these fluctuating relations.

[72] "Khronika Moskovskogo gorodskogo upravleniia," *Izvestiia Moskovskoi gorodskoi dumy*, 1911, no. 8: 23–27, 30–31; "Khronika i melkie zametki," *Elektrotekhnicheskoe delo*, 1911, no. 3: 22.

Moscow and St. Petersburg remained the premier cities of electrification through the 1920s. Their foreign-owned utilities continually introduced new equipment and techniques into Russia, trained administrative and technical staff, and pioneered industrial usage of electrical energy. The two cities also broke new administrative and legal ground, which provided guidelines for other cities. Moscow and St. Petersburg also led the move for alternate approaches to electrification. But before we examine this issue, it is instructive to trace the evolution of Baku and the other tiers.

Baku's industry and competitive environment created a unique market for electric light and, especially, power. A city of 300,000, it was the center of Russian oil production and a focus of foreign investment.[73] In less than three decades, Baku's electric power systems evolved from isolated and hazardous operations to the country's third-largest utility, with an industrial load far greater than in any other Russian city—90 percent compared with 0–50 percent elsewhere.[74]

Arc lighting of docks in 1880 and the Nobel refining facility in 1882 introduced electrical applications. By the 1890s, oil firms used electricity extensively for lighting.[75] These early stations suffered from inadequate safety precautions, untrained people, and inadequate oversight from the overburdened post and telegraph department, which supervised construction.[76] Technical and organizational improvements soon provided a lighting that proved far safer than kerosene and became standard for oil exploration and drilling.

The Benkendorf Company pioneered the use of electricity to power extraction pumps in 1897 and was quickly followed by Nobel, technologically and managerially the most advanced firm in Baku.[77] In 1901, a German concession, Elektricheskaia Sila (Electric Force), began operations with one AC station and opened a second in 1902 under the guidance of Marxist Leonid B. Krasin.[78] By 1908, nine other stations

[73] Audrey Altstadt, "Baku: Transformation of a Muslim Town," in Michael F. Hamm, ed., *The City in Late Imperial Russia* (Bloomington: Indiana University Press, 1986), 285–88.

[74] "Statisticheskie svedeniia . . . za 1914 god," *Elektrichestvo*, 1917, nos. 4–6: 98.

[75] "Obzor," *ZIRTO*, 1887, no. 3: 104; Gudrat Ia. Abdulsalimzade, *Osushchestvlenie Leninskogo plana elektrifikatsii v Azerbaidzhane* (Baku: Izdatelstvo Akademii nauk ASSR, 1968), 14–15; Robert W. Tolf, *The Russian Rockefellers: The Saga of the Nobel Family and the Russian Oil Industry* (Stanford: Hoover Institution Press, 1976), 144.

[76] E. M. Iushkin, "Ob elektricheskom osveshchenii na neftianykh promyselakh," *Elektrichestvo*, 1900, no. 14: 192.

[77] A. Beeby Thompson, *The Oil Fields of Baku and the Russian Petroleum Industry* (London: Crosby Lockwood and Son, 1908), 242.

[78] Abdulsalimzade, *Osushchestvlenie*, 16–18; Tolf, *Russian Rockefellers*, 146–47.

had entered the country's most competitive market, but their reliance on DC power limited their area of service and, ultimately, their success.[79] By January 1914, electric motors provided 37 percent of the horsepower for the oil pumps compared with 43 percent by steam engines and 20 percent by internal combustion engines.[80] This trend toward electrifying oil pumps continued through the war. Larger firms tended to electrify quicker than smaller firms because of easier access to financial and technical resources.[81] Total private and utility output grew tenfold from 18 to 187 MkWh in the decade after 1905 as oil production expanded and electric motors replaced steam generators.[82]

Of the 187 MkWh produced in 1914, Elektricheskaia Sila generated 154 MkWh (82 percent).[83] The company's two stations had grown sevenfold in capacity and fourteenfold in output since 1902, a major increase in operating efficiency. From 1905 to 1915, four expansions increased capacity from 6 to 45 MW yet were unable to satisfy demand.[84] Elektricheskaia Sila monopolized the Baku electricity supply, not just because of its more efficient stations, but also because of its aggressive, even predatory tariff policies against potential and real competitors. The firm adjusted its rates to keep large users from building their own stations and compensated by charging higher rates to smaller users who could not afford building a station.[85] The utility's 1914 industrial tariffs ranged from 4 to 15 kopecks per kWh, one of the widest spans in Russia, while its tariff for city use stayed at a high 28 kopecks per kWh until it dropped to 20.5 kopecks in May.[86] If the 1914 dividend of 9 percent was an indicator, the firm produced above average utility profits.[87]

The Second and Third Tiers, 1895–1914

As IN OTHER AREAS, the rapid growth of factory demand supplied by the first-tier utilities placed these cities in the forefront of industrial modernization. In contrast, the second- and third-tier utilities inhab-

[79] "Statisticheskie svedeniia . . . v Rossii," *Elektrichestvo*, 1910, no. 1: 16.
[80] TsGIAL f. 23, op. 28, d. 2554, 78.
[81] Azneft, *Bakinskaia Neftianaia promyshlennost za tri goda natsionalizatsii (28 maia 1920–28 maia 1923)* (Baku: Azneft, 1923), 20–21; Abdulsalimzade, *Osushchestvlenie*, 32–33.
[82] TsGIAL f. 23, op. 28, d. 2554, 45–46, 171.
[83] Ibid.; "Statisticheskie svedeniia . . . za 1914 god," *Elektrichestvo*, 1917, nos. 4–6: 98.
[84] Abdulsalimzade, *Osushchestvlenie*, 23–26, 34, 36.
[85] TsGIAL f. 23, op. 28, d. 2554, 59–60.
[86] Abdulsalimzade, *Osushchestvlenie*, 40. Only Kherson equalled the wide range; "Statisticheskie svedeniia . . . za 1914 god," *Elektrichestvo*, 1917, nos. 4–6: 56–73.
[87] TsGIAL f. 23, op. 28, d. 2554, 46.

ited a different world, one where electrification penetrated far less into the life of cities and towns. The major reasons for this delay were bureaucratic and financial, not technical. Not until the 1897 model agreement did municipalities begin to approve utilities, while the weak domestic financial markets, municipal reluctance to add debt, and lack of a supporting infrastructure continued to slow utility diffusion and expansion.

Another retarding factor was the low rate of urbanization. Only a sixth of Russia's population lived in urban areas by 1913, compared with two-fifths of Europe's population.[88] Furthermore, most of the Russian urban population lived in towns of less than 100,000 people, the great majority of which lacked factories, railroads, and other modern infrastructure.[89] Electrification demanded an adequate return from a large investment, and only urban areas had the density and customer base for profitable operations. By their greater urbanization, Europe and the United States possessed greater potential markets with a larger, more developed, and more receptive economic and industrial base for electrification.

The Moscow and St. Petersburg utilities remained alone in Russia through the 1880s, despite interest by other cities in an 1886 Company concession.[90] A few cities approved concessions in the early and mid-1890s, but the real spread of central electric stations in European Russia began after 1898 and skyrocketed after 1908. The two decades between the first concessions and widespread application highlight the problem in transferring and diffusing new technology beyond Moscow and St. Petersburg.

The lack of an administrative and legal framework was a greater hindrance initially than the high cost of constructing a utility. Not until the 1897 St. Petersburg model agreement did municipal and tsarist officials have the security of written guidelines on utility concessions. The model agreement greatly aided the spread of utilities by providing officials with an administrative foundation.[91] Further pro-

---

[88] Chauncy D. Harris, *Cities of the Soviet Union* (Washington, D.C.: Rand McNally, 1972), 232; Paul M. Hohenberg and Lynn Hollen Lees, *The Making of Urban Europe, 1000–1950* (Cambridge: Harvard University Press, 1985), 219.

[89] *The Russian Year-Book for 1911* (London: Eyre and Spottiswoode, 1911), 292; *The Russian Almanac 1919* (London: Eyre and Spottiswoode, 1919), 157; David R. Brower, "Urban Revolution in the Late Russian Empire," in Hamm, ed., *City in Late Imperial Russia*, 319–53.

[90] Including Odessa, Kiev, and Saratov; see Kamenetskii, *Pervye russkie elektrostantsii*, 29.

[91] E. R. Ulman, "Razvitie tsentralnykh elektricheskikh stantsii v Peterburge za desiatiletnyi period," *Elektrichestvo*, 1912, no. 4: 118–20.

Graph 3.1. Growth of second- and third-tier utilities, 1893–1926

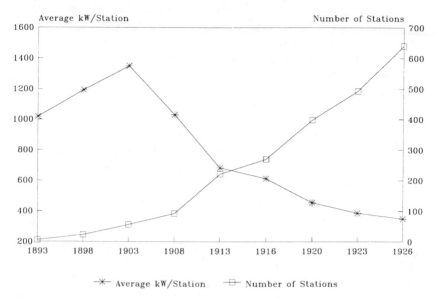

Source: "Biulleten," *Elektrichestvo*, 1927, no. 1: 43.
Note: Excludes Moscow, St. Petersburg, Baku, and regional stations.

motion came from German, Belgian, Austrian, and French manufacturing firms eager to expand their markets; European banking syndicates; and a growing domestic electrotechnical industry that supplied half of the country's equipment by 1914.[92] As Graph 3.1 demonstrates, the establishment of utility stations accelerated sharply with the economic boom after 1908. As smaller cities and towns built powerplants, however, the average station size dropped sharply. Both trends continued after World War I.

The technical characteristics of second- and third-tier utilities differed greatly from those of the first-tier stations. The major difference was the preference for DC over AC on economic and technical grounds. The strong Russian bias for DC reflected the small station size and the small geographic coverage; of 115 utilities in 1908, seventy-eight were DC, thirty-three AC, and four dual capacity.[93] A

[92] Diakin, *Germanskie kapitaly*, 25, 84; S. A. Gusev, *Razvitie sovetskoi elektrotekhnicheskoi promyshlennosti* (Moscow: Energiia, 1964), 12.
[93] "Statisticheskie svedenia . . . v Rossii," *Elektrichestvo*, 1910, no. 1: 16–18.

smaller sample in 1914 provided a similar division.[94] For most utilities, DC's lower cost, solidity of installation, simplicity of maintenance, larger financial returns, and convenient use far outweighed AC's more efficient transmission.[95]

AC use depended on large demand, which corresponded to city population, industrial use, and a need to transmit electricity over long distances.[96] AC use increased with population: Twenty-two cities with more than 100,000 people in 1912 possessed nineteen of the twenty-one AC stations.[97] The larger the population of a city, the greater its geographic spread and potential industrial and lighting base, and thereby the justification for the more expensive initial investment. Few towns and cities could immediately justify a more expensive utility. DC dominance denoted not ignorance or backwardness but conscious economic and technical choice that accurately reflected a utility's environment.

The Russian preference for DC had a European counterpart. In his study of four German towns, Edmund N. Todd has shown that local factors strongly influenced municipal choice of AC or DC.[98] Of 2,770 German electric stations in 1909, two-thirds (1,858) were DC. AC exceeded DC stations in capacity only after 1910 and in number in 1913, when 1,880 DC stations had 300 MW compared with 900 MW from 1,882 AC stations.[99] Of Great Britain's 1,360 MW of utility capacity in 1909, DC stations had 556 MW, mixed AC–DC stations 605 MW, and AC stations only 199 MW.[100] These data strongly indicate that, although the largest, most advanced stations used AC, the economic and technical attractions of DC lasted far longer than previously thought.

Because of their need for energy-dense fuels, Russian utilities had more advanced fuel patterns than the rest of the economy, which

[94] Of ninety-three utilities, sixty-two were DC, twenty-six AC, and five dual capacity; "Statisticheskie svedeniia . . . za 1914 god," *Elektrichestvo*, 1917, nos. 4–6: 55.

[95] "Deiatelnost obshchestva," *ZIRTO*, 1896, no. 3: 26.

[96] E. g., the port of Kronstadt chose 3-phase AC in 1901 "in view of the large distances over which current must be transmitted"; "Elektricheskaia stantsiia dlia Kronshtadskogo porta," *Elektrichestvo*, 1901, no. 6: 84.

[97] "Statisticheskie svedeniia . . . za 1912 god," *Elektrichestvo*, 1914, no. 6: 165–66.

[98] Edmund N. Todd, "A Tale of Three Cities: Electrification and the Structure of Choice in the Ruhr, 1886–1900," *Social Studies of Science* 17 (1987): 387–412.

[99] "Installations, Systems and Appliances," *Electrical World*, 24 February 1910, 478; "German Central-Station Statistics," *Electrical World*, 10 January 1914, 105–6. The remaining stations were mixed.

[100] "Installations, Systems and Appliances," *Electrical World*, 17 February 1910, 419.

Graph 3.2. Fuel choice in 1914, number of stations

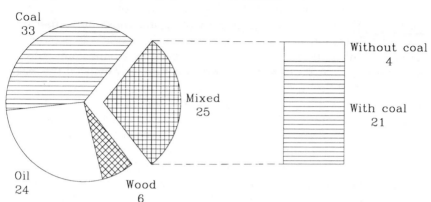

Source: "Statisticheskie svedeniia o tsentralnykh eletricheskikh stantsiiakh v Rossii za 1914 god," *Elektrichestvo*, 1917, nos. 4–6: 94–97.

depended on wood.[101] Coal and oil fueled over 90 percent of utilities (see Graph 3.2). Stations primarily burned Russian coal, mainly from the Donets basin, although utilities with access to the Baltic Sea, such as those in St. Petersburg and Riga, imported Cardiff coal. The Moscow and St. Petersburg utilities initially burned Russian coal. After tests in 1888 of British coal in St. Petersburg, the 1886 Company began regular use of Cardiff smokeless in 1891. It soon became the capital's standard, with boiler equipment designed for its use.[102] Some Polish stations, such as Lodz, burned Silesian coal. Eleven (12 percent) of ninety-three stations burned wood, four along with oil. As is not surprising, these stations were smaller than purely coal- or oil-fired stations: six produced less than 150 kW and only one exceeded 500 MW.[103]

Oil came from Baku and Grozny, source of electricity's competitor, kerosene. Russian stations were among the world's first utilities to

[101] L. K. Ramzin, "The Power Resources of Russia," in W. R. Douglas Shaw, comp. and ed., *Transactions of the First World Power Conference* (London: Percy Lund Humphries, 1924), vol. 1: 1251.

[102] "Razvitie oborudovaniia tsentralnykh elektricheskikh stantsii v Peterburge," *Elektrichestvo*, 1912, no. 6: 188; Kamenetskii, *Pervye russkie elektrostantsii*, 50.

[103] "Statisticheskie svedeniia . . . za 1914 god," *Elektrichestvo*, 1917, nos. 4–6: 74–85, 94–97.

burn oil, prompted by Nobel's efforts to increase its market.[104] The 1886 Company tested *mazut* (heating oil) in Moscow in 1889–90, one of the first utility uses of oil in the world, and chose oil to fuel the Georgievsk station in 1897.[105] Kiev opened the world's first oil-fired electric plant in 1892.[106] Oil use depended on economic competitiveness with coal. A drop in mazut prices in 1901–2 convinced the 1886 Company and Smirnov stations in St. Petersburg to convert a significant percentage of their boilers to oil.[107] The sharp rise in prices during the 1905–6 revolution triggered reconversions to coal and destroyed any hope of Russian oil supplanting British coal in the capital.

The overall problem facing cities was how to meet increasing demands for services from a rapidly expanding urban population without the financial means and legislative freedom from the state bureaucracy to satisfy them.[108] Municipal governments received guidance and information from the 1897 St. Petersburg model agreement, the VI Section, and the imperial government. The model agreement provided an invaluable legitimizing precedent for city officials with little knowledge of concessionary agreements. City governments eventually discovered that the agreement suffered from vague administrative powers, inadequate financial returns for the city, and loose formulas and definitions.[109] These shortcomings, caused as much by the reluctance to deviate from the original text as from the limited expertise available in 1897, illustrate the administrative conservatism that pervaded local as well as national government. In sharp contrast, the American approach of municipal franchises accelerated the diffusion of network technologies.[110]

---

[104] Tolf, *Russian Rockefellers*, 70, 147.

[105] Kamenetskii, *Pervye russkie elektrostantsii*, 51.

[106] Donald E. Thomas, Jr., *Diesel, Technology and Society in Industrial Germany* (Tuscaloosa: University of Alabama Press, 1987), 206.

[107] Ulman, "Razvitie tsentralnykh elektricheskikh stantsii v Peterburge za desiatiletnyi period," 118.

[108] Michael F. Hamm, "The Breakdown of Urban Modernization: A Prelude to the Revolutions of 1917," in Michael F. Hamm, ed., *The City in Russian History* (Lexington: University Press of Kentucky, 1976), 182–200; Thomas C. Owen, *Capitalism and Politics in Russia: A Social History of the Moscow Merchants, 1855–1905* (Cambridge: Cambridge University Press, 1982), 158–59.

[109] E. A. Shuks, "Doklad Podgotovitelnoi komissii po vyrabotke normalnogo kontsessionnogo dogovora," *Elektrichestvo*, 1915, no. 16: 305–7.

[110] Letty Anderson, "Fire and Disease: The Development of Water Supply Systems in New England, 1870–1900," in Joel A. Tarr and Gabriel Dupuy, eds., *Technology and the Rise of the Networked City in Europe and America* (Philadelphia: Temple University Press, 1988), 140–41.

City governments had to overcome several obstacles for electric current to flow. Besides the administrative marathon described in the previous chapter, a concession had to be granted or a contract let to build a station, financing found, and the new utility integrated into city operations. None of these steps were easy, and all demanded time. Generally, a municipality established an uprava technical commission to control the utility and a duma oversight committee. A city's request for proposals usually led to several offers, which the VI Section often evaluated. In a successful example of informal centralization to maximize availability of skilled personnel, the VI Section served as an invaluable source of technical expertise that cities often lacked.

Yalta exemplifies the process. In May 1903, the town government asked the VI Section to evaluate bids for lighting. The section's commission judged all ten submissions unsatisfactory because of inadequate data and incompatible assumptions. Finding it impossible to make direct financial comparisons among the bids, the commission redefined the technical goals and invited firms to resubmit offers. The commission recommended not the latest technology but the simplest equipment to meet present demands while leaving room for expansion.[111]

Municipalities considered the financial aspects carefully. Ultimately, a utility had to be economically viable. Securing a customer base required action on several fronts: creating a local awareness of electricity and demonstrating its superiority to kerosene, offering an economically affordable and competitive service, and providing a reliable and technically compatible product to the subscriber. The economic aspects of proposals—revenues for the city and costs to its citizens—usually proved more important to the local government than did technical aspects.[112] Besides a fixed fee or percentage of the concession's gross income, cities asked for free or discount lighting for government buildings and streets.[113]

In Russia, capital was expensive; the large labor forces of many factories indicated not so much technical backwardness as insufficient

---

[111] "Deiatelnost VI," *ZIRTO*, 1904, nos. 9–10: 261; "Zhurnal zavedanii Komissii po rassmotreniu proektov na ustroistvo elektricheskogo osveshcheniia v gorode Yalte," *ZIRTO*, 1904, no. 11: 325–30.

[112] E. g., the discussion about choosing a concession for Piatigorsk; "Po russkim gorodam," *Elektrichestvo*, 1910, no. 1: 45.

[113] E. g., in Volkhov the twelve-year concession stipulated free lighting for the city hall and 425 600-watt lights; "Munitsipalnoe obozrenie," *Gorodskoe delo*, 1912, no. 18: 1158.

capital to invest.[114] Since, unlike the manufacturing sector, utilities could not substitute labor for machinery, access to capital was important. An exception was coal handling, for which labor proved less expensive than automated equipment.[115] Cities had to float loans or otherwise secure capital approved by the Ministry of Finance and the MVD. After 1910, the imperial government greatly relaxed its process for city loans, which aided municipal ownership of utilities.[116]

A key issue for a municipality was whether it should operate the utility or give a concession to a private firm. Concessionnaires provided a package of technical expertise and financing, two ingredients of a high technology municipal governments often lacked, whereas municipalization offered more monies for city budgets.[117] As in Canada, private concessions appeared first, taking the major financial and technological risks.[118] The larger the city, the earlier it electrified, and the larger its need for capital, the more likely its utility was a concession (see Table 3.2). Foreign concessionaires dominated cities of more than 250,000 in a demonstration of superior financing, technical expertise, and experience. These firms operated as large Russian joint stock companies with shares traded on the St. Petersburg stock exchange. The trend after 1906 was toward municipal operations, discussed below.

Municipalities and utilities had, of necessity, a close if not always friendly relationship. City governments and concessions faced inherent conflicts about costs, benefits, service, coverage, and oversight.[119] The first and last years of a contract proved the most difficult. In the former, each party had to learn to work with the other in an often difficult and slow process. In the latter and especially during a man-

---

[114] Olga Crisp, *Studies in the Russian Economy before 1914* (London: Macmillan, 1976), 40–41.

[115] Ulman, "Razvitie tsentralnykh elektricheskikh stantsii v Peterburge za desiatiletnyi period," 118.

[116] Michael F. Hamm, "Khar'khov's Progressive Duma, 1910–1914: A Study in Russian Municipal Reform," *Slavic Review* 40 (March 1981): 33.

[117] Guenter S. Holzer, "German Electrical Industry in Russia: From Economic Entrepreneurship to Political Activism, 1890–1918" (Ph.D. diss., University of Nebraska, 1970), 45; J. H. Bater, "Modernization and the Municipality: Moscow and St. Petersburg on the Eve of the Great War," in James H. Bater and R. A. French, eds., *Studies in Russian Historical Geography*, vol. 2 (London: Academic Press, 1983), 320.

[118] Christopher Armstrong and H. V. Nelles, *Monopoly's Moment: The Organization and Regulation of Canadian Utilities, 1830–1930* (Philadelphia: Temple University Press, 1986), 31.

[119] See complaints about concessions in Elisavetgrad, Krasnoiarsk, and Kharkov in "Khronika gorodskikh upravlenii v Rossii," *Izvestiia Moskovskoi gorodskoi dumy*, 1907, no. 5: 68–72.

dated transfer or buyout, disputes easily developed over the details of transfer.[120] Fear of a takeover could cause a utility to reduce its investment, thus lowering the quality of service and increasing the movement for a takeover.[121] By creative use of reserve funds, accounting definitions, and other techniques, concessions tried—or so city governments believed—to reduce their contractual obligations to the municipality.[122] Such an atmosphere of mutual distrust only added to the poor public perception of utilities.

Symbolic of Russia's slow economic and technological development, the third tier did not develop either widely or deeply under tsarism. This tier marked the least and latest expansion of utilities into Russian society. According to *Elektrichestvo* statistics, the twenty-eight utilities in towns with fewer than 50,000 people in 1914 provided the least capacity and most expensive service, the consequence of a limited customer base and small, less efficient stations.[123] Two-thirds of these utilities began operations after 1905. Local governments owned sixteen stations, indicating their unattractiveness to private investors and the growing municipalization movement. DC powered 90 percent of the stations, which averaged 211 kW per station, nearly two orders of magnitude less than first-tier utilities. Lighting completely dominated demand: ten stations had no industrial load, eight had a load less than 20 percent, and only one had a load greater than 50 percent. Of twenty-three towns, fifteen had a per capita consumption below 10 kWh, seven ranged between 11 and 20 kWh, and one consumed more than 20 kWh per inhabitant. In contrast, Moscow averaged 92 kWh per capita.[124] Electric light, much less electric power, barely penetrated beyond the province of local elites into the social and economic fabric of daily life.

Not registering at all on *Elektrichestvo*'s statistics were the 101 prewar rural stations with an average size of 93 kW, half the size of the third-tier utilities.[125] As in the West, rural stations attracted little attention. Unlike those in the West, rural stations attracted much attention after 1920, when their political importance rapidly increased.

---

[120] E. g., in Nizhni-Novgorod the company threatened to stop generating power in June 1915, several months before the city-owned replacement station would be ready; "Iz gazet," *Elektrotekhnicheskoe delo*, 1914, nos. 7–8: 29.

[121] E. g., Arkhangelsk; "Po russkim gorodam," *Elektrichestvo*, 1913, no. 1: 45.

[122] "Khronika gorodskikh upravlenii v Rossii," *Izvestiia Moskovskoi gorodskoi dumy*, 1912, nos. 6–7: 36–38.

[123] "Statisticheskie svedeniia . . . za 1914 god," *Elektrichestvo*, 1917, nos. 4–6: 56–100.

[124] Ibid., 58.

[125] "Biulleten," *Elektrichestvo*, 1923, nos. 7–8: 402.

Graph 3.3. Distribution of utility generation, 1905 and 1913

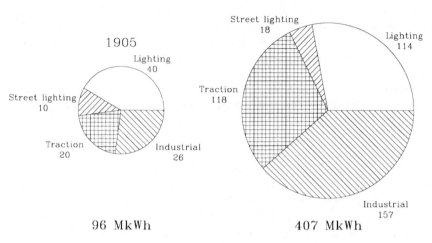

96 MkWh   407 MkWh

Source: "Zapiska VI (elektr.) otdela IRTO po voprosu ob oblozhenii aktsiz elektri-cheskoi energii, idushchei dlia tseli osveshcheniia," *Elektrichestvo*, 1915, no. 1: 21, 28.

Despite the late start, utility diffusion and output increased rapidly after 1900 as the economy expanded and electric light, power, and traction spread. Utility generation increased fourfold from 96 MkWh in 1905 to 407 MkWh in 1913, paralleling similar growth in private and industrial stations.[126] Industrial and tram consumption increased twice as much as lighting demand, a portent for future industrial modernization (see Graph 3.3).

As in the West, factories generated most of the electricity consumed by industry, but utilities increased their share of this market. The very uneven distribution of industrial load among utilities and low percentage of industrial demand supplied by utilities, consequences of the structure of Russian industry and the inability of utilities to expand rapidly, distinguished Russia. Utilities supplied a significant part of industrial demand only in first-tier and a few second-tier cities. Lighting overwhelmed industrial demand for most utilities (see Graph 3.4). In the second and third tiers, only fourteen of one hun-

[126] Data from 1905 Ministry of Finance and 1913 VI Section surveys: "Zapiska VI (elektr.) otdela IRTO po voprosu ob oblozhenii," 28; and "Khronika," *Elektrichestvo*, 1917, nos. 13–14: 178.

Graph 3.4. Utility industrial load by tier, 1914

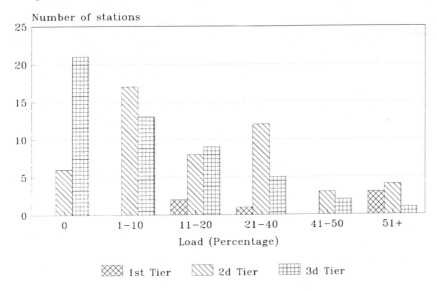

Source: "Statisticheskie svedeniia . . . za 1914 god," *Elektrichestvo*, 1917, nos. 4–6: 98–100; "Zapiska VI (elektr.) otdela IRTO po voprosu ob oblozhenii aktsiz elektricheskoi energii, idushchei dlia tseli osveshcheniia," *Elektrichestvo*, 1915, no. 1: 26–29.

dred utilities had an industrial load greater than 30 percent.[127] Indeed, in the industrial city of Ivanovo-Voznesensk, the Society of Factory Owners sold electricity to the city government.[128]

One reason for the small utility share of industrial demand was the structure of Russian industry. Factories tended to be either very large or very small.[129] The larger the plant, the more likely it operated its own electric station. The smaller the plant, the less likely it used any electric power. Medium plants, those with enough demand to warrant electric power but not enough to justify their own station, rarely existed outside the first tier in the numbers necessary for a large in-

[127] "Zapiska VI otdela IRTO," 26–29.

[128] *Ivanovo-Voznesenskaia guberniia za desiat let Oktiabrskoi revoliutsii* (Ivanovo-Voznesensk: Gorod Ivanovo-Voznesensk, 1927), 25–26.

[129] Leon Trotsky, 1905, trans. Anya Bostock (New York: Random House, 1971), 19–22; Ia. S. Rozenfeld, *Promyshlennaia politika SSSR (1917–1925 gg.)* (Moscow: Planovoe khoziaistvo, 1926), 46.

...

dustrial load. Paralleling utilities, small factories used DC and large factories used AC.[130]

As in the West, the largest utilities quickly perceived the importance of daytime industrial demand to increase their load factor, the ratio of average to maximum generation.[131] The higher the load factor, the less each unit of electricity cost to generate and the greater the plant's economic and technical efficiency. Since the maximum demand on a utility was usually evening lighting, stations had unused daytime capacity. Utilities promoted differential tariffs to lure industries, whose peak loads differed from those of residential and business users in time, and pushed the development of other types of consumption, including heating, cooling, and ventilation.[132] Consequently, industrial users usually paid significantly less for daytime consumption than did individuals for evening lighting.

Industrial electrification concentrated in Baku, Moscow, and St. Petersburg, reflecting the geographic concentration of manufacturing. The 1886 Company in Moscow made a concerted attempt to increase its industrial load by adding capacity, expanding into the city's industrial outskirts, adopting a differential tariff in 1907, and reducing rates 20 percent in 1912. From 1903 to 1913, 286 firms switched to the company; more than half converted after 1909.[133] Of the eleven major cities of the Central Industrial Region around Moscow in 1914, industrial users consumed more than 3 percent of utility output in three cities and in only one did they consume more—10 percent in Smolensk.[134]

The low industrial load of second- and third-tier utilities did not lessen their inability to meet demand. A survey of sixty cities in electrical and urban journals after 1908 shows increased diffusion, quick growth, good profits, more consumers, and greater output. Consumption grew rapidly, forcing expansion and new investment along with large gross profits and dividends averaging 5–10 percent. As

[130] K. A. Krug, *Elektrifikatsiia Tsentralno-promyshlennogo raiona* (Moscow: Teplovoi komitet pri Politekhnicheskom obshchestve no. 7, 1918), 34–35.

[131] Hughes, *Networks of Power*, 218–21; Harold L. Platt, "City Lights: The Electrification of the Chicago Region, 1880–1930," in Tarr and Dupuy, eds., *Rise of the Networked City*, 261–64.

[132] M. A. Grinberg, "O tarifakh tsentralnykh elektricheskikh stantsii," *Gorodskoe delo*, 1912, no. 17: 1069–75. Grinberg noted the numerous American pamphlets on promotion of offpeak use.

[133] V. D. Kirpichnikov, "Razvitie Moskovskoi tsentralnoi elektricheskoi stantsii O-va 1886 goda," *Elektrichestvo*, 1914, no. 3: 86; "Iz gazet," *Elektricheskoe delo*, 1914, no. 12: 19–20.

[134] Krug, *Elektrifikatsiia Tsentralno-promyshlennogo raiona*, 30.

Graph 3.5. Kharkov's utility capacity, 1897–1914

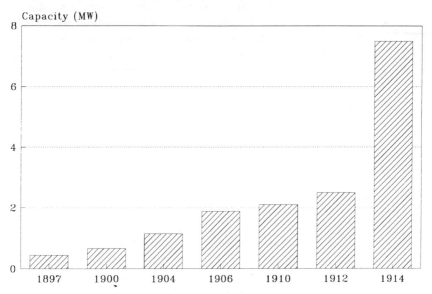

Sources: "Soobshcheniia s mest," *Gorodskoe delo*, 1911, nos. 13–14: 1032; "Gorodskaia elektricheskaia stantsiia gor. Kharkova," *Elektrichestvo*, 1915, no. 8: 140–41; 1912: "Statisticheskie svedeniia o tsentralnykh elektricheskikh stantsiiakh v Rossii za 1912 god," *Elektrichestvo*, 1914, no. 6: 187; 1914: "Statisticheskie svedeniia o tsentralnykh elektricheskikh stantsiiakh v Rossii za 1914 god," *Elektrichestvo*, 1917, nos. 4–6: 77.

systems reached saturation, utilities had to invest in expanding capacity. A wider base of users meant more business but also more outlays of capital in a never-ending circle as utilities sought funding to enlarge existing stations.[135] Obtaining the necessary permissions from the tsarist bureaucracy, however, demanded much time. This problem, along with difficulties in obtaining financing, ensured that capacity continually lagged behind demand. One frequent consequence was a moratorium on new users until a utility expanded, which could take years.[136] Kharkov, a city of a quarter-million, exemplifies these issues. Its utility developed early, stagnated for half a decade, and then grew again, switching from DC to AC in the process (see Graph 3.5). Its municipal station opened in 1897 with a 440-kW DC system and increased capacity fourfold in a decade. In 1906, the addition of

[135] E. g., Viatka; "Iz gazet," *Elektricheskoe delo*, 1911, no. 6: 22.
[136] E. g., Rostov-on-Don; "Iz gazet," *Elektricheskoe delo*, 1911, no. 7: 24.

an electric tram saturated the station load, and the city stopped adding new customers for six years.[137] As with utilities elsewhere, capacity was reached in lighting, not in industrial consumption. Kharkov's
load was 80 percent lighting, 10 percent tram, and 10 percent industrial.[138] Faced with a lack of capacity and high tariffs (40 kopecks per
kWh for home lighting, 30 kopecks for commercial lighting, and 20
kopecks for industrial use), several factories and other users built
their own stations, even taking on customers. A separate tram station
opened in 1910.[139] The city finally received a loan in 1911 to construct
a new 3-phase AC central station, which opened in 1912 and incorporated the tram station, and a loan in 1914 to construct an AC distribution network.[140] The industrial tariff dropped in 1913 to 15 kopecks to
increase the load.[141] Consumption per person grew from 10 kWh in
1909 to 20 kWh in 1912 and 30 kWh in 1914 as consumption quickly
absorbed new capacity.[142]

Dividends and gross profits varied over time and by city, but they
generally followed the fortunes of the economy. The median dividend
fluctuated between 5 and 8 percent, slightly higher than most foreign
profits, which ranged from 4 to 9 percent between 1890 and 1914.[143]
The recession of 1899–1902 hurt utility revenues and slowed the establishment of new stations. The 1886 Company dividends in 1900
and 1901 sank to 1 percent and 1.5 percent, respectively.[144] The
post-1908 economic expansion similarly benefited utilities.[145]

The cost of electricity fell sharply over time but remained relatively
expensive compared with the West. In Moscow, a kilowatt-hour in
1887 cost 50 or 65 kopecks, depending on the use. In 1899, a kilowatt-
hour cost 40 kopecks compared with a range in 1912 from 5 to 20

[137] Smolensk experienced a similar delay; see I. Shirman, "Elektricheskoe osveshchenie i tramvai v g. Smolenske," Elektrichestvo, 1917, nos. 2–3: 40.
[138] "Gorodskaia elektricheskaia stantsiia gor. Kharkova," Elektrichestvo, 1915, no. 8:
140–41.
[139] "Soobshcheniia s mest," Gorodskoe delo, 1911, nos. 13–14: 1032.
[140] "Iz gazet," Elektricheskoe delo, 1912, no. 1: 22; 1914, no. 4: 22.
[141] "Iz gazet," Elektricheskoe delo, 1913, no. 12: 22.
[142] For 1909, see "Gorodskaia elektricheskaia stantsiia gor. Kharkova," Elektrichestvo,
1915, no. 8: 140; for 1912, see "Statisticheskie svedeniia . . . za 1912 god," Elektrichestvo,
1914, no. 6: 176; for 1914, see "Statisticheskie svedeniia . . . za 1914 god," Elektrichestvo,
1917, nos. 4–6: 60.
[143] P. V. Ol, Foreign Capital in Russia (New York: Garland, 1983), 251.
[144] "Elektrotekhnika v Rossii," Elektrotekhnicheskii vestnik, 1902, no. 23: 550.
[145] E. g., the municipal utility of Irkutsk reported a 10 percent gross profit in 1911, its
first year, and gross profits of 35–40 percent for the next three years; "Po russkim
gorodam," Elektrichestvo, 1915, no. 12: 263–64.

kopecks.[146] A kilowatt-hour in Moscow or St. Petersburg cost two to three times more than in Berlin.[147]

Moscow and St. Petersburg had the country's lowest tariffs and wealthiest populations, a combination that produced the highest per capita consumption. Tariffs and cost elsewhere differed greatly by type and location. In 1912, tariffs ranged from 5 to 50 kopecks per kWh with discounts for quantity, time of day, and form of consumption.[148] In 1913, actual costs ranged from 2 to 40 kopecks per kWh.[149] The wide range of costs matched utility size. Because small stations had lower operating efficiencies and higher expenses, generating electricity cost them more than it did the 1886 Company.

Although imprecise, data on per capita consumption indicate a low national level of electrification, but, as with tariffs, wide variations existed. The higher the consumption per person, the more electric light and power had penetrated daily life. The per capita consumption in eighty-three cities in 1914 falls into two distinct groups of low and high consumption, as Graph 3.6 illustrates. Fifty cities averaged between 6 and 20 kWh per person and another twenty-three consumed less than 6 kWh per person. All these cities were in the second and third tiers, as were five cities with 21–40 kWh per person. Only five cities—Moscow, St. Petersburg, Baku, Lodz, and Bogorod—averaged more than 60 kWh per person, approximately seven times more than the vast majority of cities.[150] As one indication of the gap between Russia and the West, only the per capita consumption of the first-tier cities came close to that of major Western cities.[151]

The lower per capita consumption compared with the West is not unexpected and confirms the backward and economically underdeveloped Russian city suggested by Michael Hamm.[152] What is unexpected is the sharp break instead of a continuum, further evidence of

[146] Kirpichnikov, "Razvitie Moskovskoi tsentralnoi elektricheskoi stantsii," 81, 83.

[147] In 1909–10, the cost of a kilowatt-hour in pfennings was Amsterdam, 25; Berlin, 40; Vienna, 60; St. Petersburg, 75.6; and Moscow, 108. O. G. Flekkel, "'Ekonomicheskie' lampochki i populiarizatsiia elektrichestva," *Gorodskoe delo*, 1912, no. 4: 238.

[148] "Statisticheskie svedeniia . . . za 1912 god," *Elektrichestvo*, 1914, no. 6: 169–81.

[149] I. A. Skavani, "K voprosu o sebestoimosti i tarifikatsii energii na russkikh tsentralnykh elektrostantsiiakh," *Elektrichestvo*, 1924, no. 4: 197–98.

[150] "Zapiska VI otdela IRTO" 26–29; "Statisticheskie svedeniia . . . za 1914 god," *Elektrichestvo*, 1917, nos. 4–6: 98–102.

[151] E. g., 110 kilowatt-hours in London, 170 in Berlin, and 310 in Chicago in 1910–13; see P. Gurevich, "Osnovnye voprosy elektricheskoi politiki v poslevoennuiu epokhu v Rossii," *Elektrichestvo*, 1917, no. 1: 10.

[152] Hamm, "Khar'khov's Progressive Duma," 32.

Graph 3.6. Russian consumption per capita, 1914

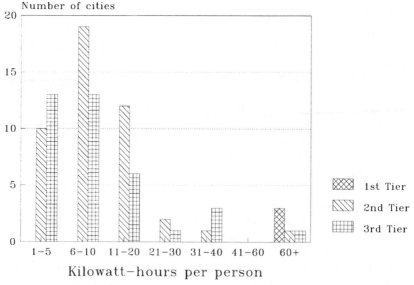

Source: "Statisticheskie svedeniia . . . za 1914 god," *Elektrichestvo*, 1917, nos. 4–6: 98–102; "Zapiska VI (elektr.) otdela IRTO po voprosu ob oblozhenii aktsiz elektricheskoi energii, idushchei dlia tseli osveshcheniia," *Elektrichestvo*, 1915, no. 1: 26–29.

the gap between the first tier and other utilities. The majority of Russian cities possessed utilities, but the benefits of electric light and power rarely reached extensively outside the first tier.

Electric Traction

ELECTRIC TRACTION was an integral part of the modern city of 1900. Compared with horse-drawn trams, electric trams lowered fares and provided better service, yet, by greatly increasing ridership and reducing operating costs, they also filled city coffers in Russia and the West. Like the electric light, trams served as both a visible sign of development and a vital component of the infrastructure that modernized Russia.

Commercial electric trams date from Frank Sprague's Richmond, Virginia, trials in the mid-1880s, but a decade of worldwide development and experimentation preceded his success. The simultaneous development of electric trams in several countries indicated the ripeness of the concept; the decade littered with numerous experiments demonstrated the work needed to forge a technically, economically, and politically feasible system.[153] Entrepreneurs had to supply electricity to a reliable motor that could handle variable speeds and frequent stops, find financing, secure local approval, and fend off competitors.[154] In the United States, large-scale commercial use started in 1888 with Sprague's operations and expanded rapidly. In Europe, the profusion of different technical systems and greater concern about the aesthetics of the overhead wires and supporting poles inhibited expansion until the early 1890s.[155]

Russian exploitation of electric trams lagged behind European development by ten to twenty years: in 1898, Germany had 73 trolley systems, France 56, Britain 29, and Russia 7. Kiev, whose steep hills made horse trams impractical, inaugurated electric traction in 1892.[156] Other cities did not convert tramways from horse to electric power or construct new electric trams until the 1900s. By 1914, forty-one Russian cities had electric trams, only one-eighth the American total of two decades earlier.[157] As in other areas of electrification, Russia had fewer systems and an average usage one-third to one-half the European level. A scornful attitude by Russian banks, the lack of technical entrepreneur-advocates, mild interest by cities, and high construction costs hindered the adoption of electric trams, and the lower degree of urbanization limited the potential market.[158]

Despite these obstacles, trams became the backbone of Russian urban public transport. The greatest growth occurred between 1900 and 1910, when track increased from 421 to 1,882 kilometers.[159] The expan-

[153] John P. McKay, "Comparative Perspectives on Transit in Europe and the United States, 1850–1914," in Tarr and Dupuy, eds., *Rise of the Networked City*, 8.

[154] Passer, *Electrical Manufacturers*, 216–18.

[155] For North America and Europe, see John P. McKay, *Tramways and Trolleys: The Rise of Urban Mass Transport in Europe* (Princeton: Princeton University Press, 1976); Sam Bass Warner, *Streetcar Suburbs* (Cambridge: Harvard University Press, 1962); Armstrong and Nelles, *Monopoly's Moment*, 85–89.

[156] P. V. Barabanov, "Pervyi elektricheskii tramvai v Rossii," *Pochtovo-telegraficheskii zhurnal*, 1893, no. 5: 466–82; no. 6: 549–63; no. 8: 703–12; no. 11: 997–1004.

[157] For Europe, see McKay, *Tramways and Trolleys*, 76. For Russia in 1898, see Martin Crouch, "Problems of Soviet Urban Transport," *Soviet Studies* 31 (April 1979): 232; in 1914, B. Veselovskii, "Munitsipalizatsiia i munitsipalnaia politika v Rossii," *ZIRTO*, 1914, no. 4: 85. For the United States in 1892, see Barabanov, "Pervyi elektricheskii tramvai v Rossii," 466.

[158] V. Shelgunov, "Belgiiskii tramvai v Rossii," *Izvestiia Moskovskoi gorodskoi dumy*, 1914, no. 3: 83; A. Vulf, "Usloviia dokhodnosti elektricheskogo tramvaia v nebolshikh russkikh gorodakh," *Gorodskoe delo* 1909, no. 4: 148.

[159] "Naruzhnoe osveshchenie," *Entsiklopediia mestnogo upravleniia i khoziaistva*, 581.

sion consisted of the conversion of horse trams to electricity, the establishment of new systems, construction of new lines, and increased utilization of capacity. Consumption of electricity by trams grew sixfold from 1905 to 1913, expanding from 20 percent (20 MkWh) to 29 percent (118 MkWh) of all utility generation.[160] Dedicated stations, using low-frequency, high-voltage DC, powered the first electric trams. About half of the later tram stations used AC and several added industrial and lighting customers for better load patterns.[161]

The slower deployment of trams and the absence of a domestic industry allowed the near monopolization by Belgian firms, which offered Russia technically mature systems accompanied by managerial expertise and financial backing.[162] In 1913, Belgian concessions operated twenty electric trams, half the country's electric trams, plus one steam and two horse trams.[163] Belgian-based funds constituted three-quarters (68 million rubles) of the 94 million rubles invested in trams between 1909 and 1914.[164] This domination continued Belgium's international entrepreneurial involvement in horse trams.[165]

Because of this Belgian presence, concessions outnumbered municipal tram operations by three to one (see Table 3.4).[166] The two significant exceptions were Moscow and St. Petersburg. After several years of municipal inaction in the late 1890s and a struggle between the ministries of finance and internal affairs over profits and ownership, the Moscow government began operating a municipal tram in 1903 and bought out the Belgian concession in 1911.[167] Following an initial city decision to minimize its investment and risk by allowing a concessionary competition in St. Petersburg, Westinghouse obtained the tram concession in 1906 but soon transferred ownership to the city government.[168] St. Petersburg also had municipally owned steam and horse trams, as well as other public transportation.[169] By 1912–14, sat-

---

[160] "Zapiska VI otdela IRTO," 21, 28.

[161] A. Vulf, "Usloviia dokhodnosti elektricheskogo tramvaia," 152–54; "Statisticheskie svedeniia . . . za 1914 god," *Elektrichestvo*, 1917, nos. 4–6: 54–73.

[162] V. I. K. "Belgiiskie kapitaly v Rossii," *ZIRTO*, 1909, no. 5: 154.

[163] Shelgunov, "Belgiiskie tramvai," 71.

[164] Diakin, *Germanskie kapitaly*, 268–69.

[165] The Brussels stock market listed 111 tram companies in 1903 and 188 a decade later, including nineteen Russian companies; neither the St. Petersburg nor the Moscow stock market listed a single tram company in 1913; see Shelgunov, "Belgiiskii tramvai," 83.

[166] Veselovskii, "Munitsipalizatsiia i munitsipalnaia politika," 85.

[167] "Khronika gorodskogo dela," *Gorodskoe delo*, 1915, no. 6: 337–38.

[168] "Business Notes," *Electrical Review*, 13 July 1906, 78; and 7 September 1906, 374; Diakin, *Germanskie kapitaly*, 51–57.

[169] James Bater, *St. Petersburg: Industrialization and Change* (Montreal: Edward Arnold, 1976), 270.

Table 3.4. Ownership of Russian trams, 1914

Type	Public[a]	Private[a]	Total[b]
Electric	9 (22)	32 (78)	41 (56)
Horse	5 (19)	21 (81)	26 (36)
Steam	4 (66)	2 (33)	6 (8)
	18 (25)	55 (75)	73

Source: B. Veselovskii, "Munitsipalizatsiia i munitsipalnaia politika v Rossii," *ZIRTO*, 1914, no. 4: 85.
[a] Percentage of type in parentheses.
[b] Percentage of total in parentheses.

uration of capacity in the central areas of Moscow and St. Petersburg prompted growing discussion about subway construction.[170]

Concessionaires and municipalities sought trams for their large profits. In Moscow the tram provided more than 20 percent of city income in 1913 (4 million rubles), and in St. Petersburg income from trams increased from 880,000 rubles in 1908 to 4.8 million rubles in 1911, two-thirds of all profits from city enterprises.[171] Russo-Belgian firms averaged dividends of 6 to 7 percent, slightly greater than other European concessions and similar to the dividends of Russian electric utilities.[172]

Trams aided the development of electrification in Russia, as in the West, by providing a market for industry, a service to city citizens, and revenues to local governments. The service and revenues allowed cities to expand geographically and changed the image of electric energy from a luxury to a necessity. The slower diffusion of trams compared with the West parallels the diffusion of utilities in Russia. A major difference was Belgian instead of German domination and municipal ownership of the largest systems. The large Belgian presence, however, reflected that country's worldwide domination of a specific technology and ability to harness international capital.

[170] "Iz gazet," *Elektricheskoe delo*, 1914, no. 4: 22; Robert W. Thurston, *Liberal City, Conservative State: Moscow and Russia's Urban Crisis, 1906–1914* (New York: Oxford University Press, 1987), 147–49.

[171] For Moscow, see Veselovskii, "Munitsipalizatsiia i munitsipalnaia politika," 85; for St. Petersburg, see "Doklad o sposobakh uluchsheniia finansov goroda S. Peterburga," *Izvestiia S.-Peterburgskoi gorodskoi dumy*, 1914, no. 1: 16. Water, slaughterhouse, and gas factories were the other important city enterprises.

[172] "Po russkim gorodam," *Elektrichestvo*, 1910, no. 1: 44; "Iz gazet," *Elektricheskoe delo*, 1912–13, no. 2: 23; *Aktsionerno-paevye predpriiatiia Rossii* (Moscow, 1912); McKay, *Tramways and Trolleys*, 159.

The Rise of the Regional Station

DISCUSSION ABOUT the future of utilities appeared publicly after 1910. Two quite different visions evolved from a common postulate of very large stations. The first approach viewed electrification as a tool for social and economic change. The second focused on providing new sources of electric power to meet the rapidly growing demand of Moscow and St. Petersburg. Although they diverged more than they converged, these two perspectives directed the thrust of mainstream electrical engineering thought. These prewar concepts of regional stations lay the groundwork for wartime proposals and the 1920 GOELRO plan of state electrification, which became a key issue in postwar debates.

The idea of regional stations was both conventional and radical.[173] Utility interest in regional stations centered in Moscow and St. Petersburg, the two cities with the most urgent technical and political demands. Utilities intended to preserve their monopolies by increasing capacity and lowering costs through economies of scale to overcome inadequate supply and discourage potential competition. Advocates of electrification for social change, such as I. Ia. Perelman, thought about reaching larger segments of the population and larger geographic areas, such as the Central Industrial Region around Moscow. They wanted regionally centralized generation to minimize costs and decentralized consumption to maximize the number of consumers.

Both individual engineers and the 1886 Company wanted to move the power plant from the city to a distant site to produce inexpensive electricity. The key enabling technologies were long-distance transmission, use of local fuels, and hydropower. With the extension of the range of transmission lines from 10–20 kilometers to 100–200 kilometers, powerplants moved from a local to a regional issue. New stations would be built next to a river for cooling and transportation or at a coal mine, peat bog, or waterfall. Patterns of fuel consumption would change as local fuels, instead of coal and oil transported by railroad, powered stations: "Naturally, there will come a time when railroads will not serve electric stations but vice-versa," wrote one enthusiastic engineer.[174] Both lines of thought viewed regional stations

---

[173] Thomas P. Hughes, "The Evolution of Large Technological Systems," in Wiebe E. Bijker, Thomas P. Hughes, and Trevor J. Pinch, eds., *The Social Construction of Technological Systems* (Cambridge: MIT Press, 1987), 57.

[174] P. V. Avtsyn, *K voprosu ustroistva i ekspluatatsii oblastnykh, raionnykh elektricheskikh stantsii v Rossii* (Moscow, 1915), 10.

as inaugurating a new era of electrification. The unit of analysis was no longer technically and economically confined to the city; instead, visions grew to encompass entire provinces and even more. After 1910, the utilities of Moscow and St. Petersburg began to restructure themselves to benefit from these proposed regional stations. Four related factors accounted for this shift to new and costly technologies: rapidly rising demand, threats to the cost and integrity of the fuel supply, the threat of competition, and fear of municipal buyouts.

Rising consumption increasingly came from suburban industrial demand and the geographic extension of tram lines, and not just in the traditional city center. Rapid industrialization made coal the largest freight on railroads after 1905—and created a concomitant dependence on the railroads to deliver the fuel.[175] The railroads, however, failed to meet this challenge because of inadequate capacity and poor management.[176] The fuel crisis of 1910–13 forced utilities to question the security and affordability of their fuel supply. Even though domestic production increased by half and imports nearly doubled, the price of Donets coal increased by half and Baku oil rose threefold.[177] As a result, utilities started considering alternative fuel sources.

The threat of competition grew as utilities extended into areas without concessions and other firms planned to build stations, sparked by high tariffs and the utilities' failure to meet new demand. The Belgian-based Company for Electric Central Stations of Fain obtained a sixty-year Moscow concession for suburban traction in 1911, but it also planned to supply electricity to other users, thereby threatening the future expansion and revenues of the 1886 Company.[178] The 1911 Fain concession probably prompted the 1886 Company's 1912 rate reduction and accelerated the construction of Elektroperedacha, discussed below.

A fourth factor was the very real fear of municipal takeovers. Under the 1897 model agreement, the St. Petersburg government had the right to purchase the 1886 Company plant in 1917. The city appeared

---

[175] Alfred J. Rieber, *Merchants and Entrepreneurs in Imperial Russia* (Chapel Hill: University of North Carolina Press, 1982), 220.

[176] J. N. Westwood, *A History of Russian Railways* (London: George Allen and Unwin, 1964), 131.

[177] Arkadii L. Sidorov, *Ekonomicheskoe polozhenie Rossii v gody pervoi mirovoi voiny* (Moscow: Nauka, 1973), 502–3, 509; *Electrical World*, 23 February 1912, 291.

[178] "Munitsipalnoe obozrenie," *Gorodskoe delo*, 1911, no. 18: 1354–55; "Khronika Moskovskogo gorodskogo upravleniia," *Izvestiia Moskovskoi gorodskoi dumy*, 1911, no. 9: 16–20; Avtsyn, *K voprosu ustroistva i ekspluatatsii*, 11, 17–18; Diakin, *Germanskie kapitaly*, 171–73.

ready to exercise its option for all utilities.[179] In short, the 1886 Company and the other St. Petersburg utilities had to respond to a more hostile environment or lose everything. The Moscow section of the 1886 Company faced similar challenges.[180]

The Moscow and St. Petersburg utilities bore the brunt of the new problems. The other first-tier city, Baku, escaped because of its abundant fuel supply and entrenched monopoly. Only one other city, Rostov-on-Don, faced the threat of an independent group building a station to electrify coal mines.[181]

Alternative fuels promised utilities independence from expensive oil and coal, as well as from those who produced these fuels. New attention focused on peat and hydropower. With the former, engineers sought to make a peasant fuel supply cities, and peat did fuel the first regional station outside Moscow. With the latter, promoters sought to bring Russia into the hydroelectric era. For St. Petersburg, hydroelectric power would harness northern rivers. Hydropower plans had been proposed since 1894, but the St. Petersburg utilities and their foreign and domestic bankers did not become interested until 1911. Unlike peat, hydropower remained on paper under tsarism.

Interest in peat derived not from its sterling qualities but from its abundance and location. With less than half the thermal value of Donets coal (3,000 vs. 7,000 calories/kilogram), peat suffered from low energy density, high water content, and difficult harvesting, making it economically feasible only for local use. Peat, consumed mostly by peasant households, provided about 1 percent of total Russian prewar fuel consumption, compared with 15 percent for oil, 22 percent for coal, and 62 percent for wood.[182] In the peat-rich Central Industrial Region, however, peat supplied 13 percent of all fuel, an order of magnitude more than the national average.[183] Russian interest in peat for industrial consumption had increased since 1900, but utilities did

[179] "Doklad gorodskoi upravy," *Izvestiia S.-Peterburgskoi gorodskoi dumy*, 1914, no. 9: 2476; "Khronika gorodskikh upravlenii v Rossii," *Izvestiia Moskovskoi gorodskoi dumy*, 1915, no. 4: 64–68.

[180] "Khronika Moskovskogo gorodskogo upravleniia," *Izvestiia Moskovskoi gorodskoi dumy*, 1911, no. 8: 30–31.

[181] "Iz gazet," *Elektrotekhnicheskoe delo*, 1911, no. 9: 24; "Munitsipalnoe obozrenie," *Gorodskoe delo*, 1914, no. 9: 572.

[182] Ramzin, "Power Resources of Russia," 1286.

[183] "Khronika i melkie zemetki," *Elektrotekhnicheskoe delo*, 1914, no. 1: 21; "Iz gazet," *Elektricheskoe delo*, 1915, no. 4: 22.

not become interested until 1911, the year the state sponsored a trip to examine peat utilization in Germany, Sweden, and Holland.[184]

In November 1911, a team operating under Robert Klasson, head of the 1886 Company's Moscow section, surveyed the Bogorod peat bog 70 kilometers from Moscow to site a novel peat-fired station. Klasson then traveled to Berlin to seek financial backing from German and Swiss banks. In January 1912, the 1886 Company took a quarter share in this consortium, which officially became the independent firm Elektroperedacha in May 1913. The company completed the station in 1914, but political disputes delayed full operations until 1915.[185] Although funded and partially inspired from abroad, Elektroperedacha was constructed by Russians using technology developed in Russia as well as foreign equipment. The firm achieved several Russian technological firsts, including the first regional station, the first large-scale harvesting and processing of peat, and the first 35- and 70-kV high-voltage transmission lines.[186]

Elektroperedacha broke new ground institutionally as well. Obtaining permission from local and national authorities to build and operate the transmission lines proved the venture's most difficult task.[187] Its political battles with local and Moscow authorities illustrated the hazards of proceeding without the aegis of an authorizing national right-of-way law. The Bogorod district *zemstvo* (local rural government) and the 1886 Company took a year to agree on a contract.[188] Disputes among Elektroperedacha, the zemstvo, and the gubernator over siting and responsibility for the transmission poles further delayed construction.[189] Consequently, Elektroperedacha connected to the Moscow network only in October 1914. The first experimental

---

[184] E. g., N. Reikel, "Torf i torfianoi koka, kak toplivo dlia parovykh kotlov," *ZIRTO*, 1902, no. 5: 257–336, and S. Bogdanov, "Primenie polvidnogo torfa k otopleniiu gravdskikh pechei i parovykh kotlov," *ZIRTO*, 1906, no. 4: 257–300; "Khronika i melkie zemetki," *Elektrotekhnicheskoe delo*, 1914, no. 1: 21.

[185] "Peat Fuel in Russia," *Electrical Review*, 23 February 1912, 291; Kamenetskii, *Klasson*, 95–104.

[186] Urban distribution networks used 6 kV. The previous high was the 20-kV lines of Baku's Elektricheskaia Sila; see Abdulsalimzade, *Osushchestvlenie*, 25. German engineers first used 100-kV lines in 1912; see A. Menge, "Distribution of Electrical Energy in Germany with Special Reference to the 'Bayernwerk'," in Shaw, comp. and ed., *Transactions of the First World Power Conference*, vol. 3: 528.

[187] Kamenetskii, *Klasson*, 105.

[188] For the drafts, see TsGIAMO f. 722, op. 1, ed. kh. 876, 1–5, 48–57, 118–27, 162–71.

[189] TsGIAMO f. 723, op. 1, ed. kh. 132, 2; Kamenetskii, *Klasson*, 105–8.

transmissions with its three 5-MW turbines followed two months later.

Full service to Moscow did not begin immediately because the duma restricted Elektroperedacha's access to the city distribution network until the company shared its profits. The city suspected that Elektroperedacha's purpose was as much to reduce the city's oversight and income from the 1886 Company as to provide less expensive electricity. The restriction sparked an angry protest from the MTP about the necessity to increase wartime output. Only in late 1915 did the duma allow the 1886 Company to accept unlimited power from Elektroperedacha, which supplied 20 percent of the city's electricity during the war.[190]

The 1886 Company planned stations like Elektroperedacha for Moscow, St. Petersburg, and Lodz to control fuel prices.[191] The Main Administration for Zemstvos encouraged this effort, assuming that it would promote peat use by other industrial users.[192] One such user, the Society of Factory and Workshop Owners of the Moscow Region, tried to build a peat-fired regional station in early 1914 for its members, but it failed to obtain the legal condemnation and compensation of land for transmission lines.[193]

Next to the light bulb taking back the night, the most vivid image of electrification is the dam, taming the awesome power of nature to benefit humanity. More prosaically, hydroelectric power rested on two technologies: water turbines to convert the force of moving water into electricity, and long-distance AC transmission to transfer the power to a distribution grid. After the successful harnessing of Niagara Falls in 1895, hydropower grew rapidly in Canada, Scandinavia, and other geographically suitable regions—with one significant exception: only two commercial hydroelectric stations existed in prewar Russia. Ignorance was not the problem; industrial hydrostations operated in Russia and its province of Finland. Nor was insufficient effort to blame; entrepreneurs submitted over a dozen proposals for St. Petersburg from 1894 through 1917. The obstacle was a lack of interest by utilities, the St. Petersburg government, and the national government.

[190] "Iz gazet," *Elektrotekhnicheskoe delo*, 1915, no. 10: 19; S. P. Stafrin, "Rabota elektricheskikh stantsii Moskovskogo raiona za 1922 g.," *Elektrichestvo*, 1922, no. 1: 36; Diakin, *Germanskie kapitaly*, 175–76, 259; Kamenetskii, *Klasson*, 105–8; Lipenskii, *Moskovskaia Energeticheskaia*, 20–23.

[191] TsGIAL f. 426, op. 1, d. 1202, 1, 3.

[192] Ibid., 21.

[193] "Iz gazet," *Elektrotekhnicheskoe delo*, 1914, nos. 7–8: 30.

Prerevolutionary hydropower activities centered on the mountainous Caucasus region and St. Petersburg, close to Finnish rivers and dependent on British coal. Industrial users introduced hydropower to Russia. In 1888, a firm unsuccessfully proposed to Terek, a city in the Caucasus, to use the Terek River for lighting and a tram.[194] In 1895, Chikolev and Klasson built the first industrial hydrostation for the army's Okhtensk gunpowder factory.[195] The Ministry of Finance and the army considered a hydrostation on the Dniepr River in the Ukraine in 1900, ultimately built three decades later under Stalin's first five-year plan.[196] By 1900, Georgia and Siberia's Lena gold mines possessed three 3-phase AC industrial hydrostations.[197] Hydroelectric utilities provided 455 kW to Piatigorsk in 1903 and 135 kW to Sukhum in 1909.[198]

In Georgia, Charles Stuart received a seventy-five-year concession for two 40-MW hydrostations in 1912 after three years of negotiations.[199] The Englishman succeeded because the distances were small and the right-of-way for the transmission lines was resolved privately. Nonetheless, implementation did not occur because the scheme was overambitious: a 40-MW station would be the fourth largest station in prewar Russia and far surpass all stations in Georgia. Other prerevolutionary proposals for the Caucasus also failed, as did proposals to harness the Dniepr River.[200]

St. Petersburg attracted the most interest and least results. The inability to create a legal framework permitting the use of waterfalls and right-of-way constituted the key bottleneck. Without such laws, companies could neither build a hydrostation nor construct long-distance transmission lines. The exception, Elektroperedacha, encountered great difficulty negotiating the placement of its lines. Although initial disinterest by utilities and investors resulted in lack of a political constituency, differences among government ministries proved the most serious barrier.

---

[194] "Khronika," *Elektrotekhnik*, 1897–98, no. 22: 1322–23.

[195] R. E. Klasson, "Elektricheskaia peredacha sily trekhfaznymi tokami na Okhtinskikh porokhovykh zavodakh bliz Peterburga," *Elektrichestvo*, 1897, no. 19: 257–67.

[196] "Utilizing Russian River Power Electrically," *Electrical World*, 25 August 1900, 278.

[197] V. L. Gvozdetskii, "Gidroelektrostantsiia 'Belyi Ugol'," in *Pamiatniki nauki i tekhniki, 1982–83* (Moscow: Nauka, 1984), 75–76. For another station, see B. A. Bakhmetev, *Gidroelektricheskaia ustanovka na reke 'Satke'* (St. Petersburg: M. M. Gitzats, 1911).

[198] Gvozdetskii, "Belyi Ugol," 75, 77; "Po russkim gorodam," *Elektrichestvo*, 1910, no. 2: 75.

[199] "Po russkim gorodam," *Elektrichestvo*, 1913, no. 1: 45, "Iz gazet," *Elektricheskoe delo*, 1913, no. 12: 23.

[200] Abdulsalimzade, *Osushchestvlenie*, 33, 46; Anne D. Rassweiler, *The Generation of Power: The History of Dnieprstroi* (New York: Oxford University Press, 1988), 22–23.

Strong incentives existed for hydropower in St. Petersburg: the capital was the largest consumer of electricity in the country, its coal and oil were shipped over great distances, and nearby rivers had substantial hydroelectric potential. Four rivers within 200 kilometers—the Narva, Imatra, Vuoks, and Volkhov—received the most attention. The first proposals surfaced in the mid-1890s, but proposals by large, well-financed, and internationally supported firms did not appear before 1910. The early proposals were first-generation efforts, launched without adequate financial and political support. The later proposals came from international consortia with solid Russian roots. Neither wave of proposals received tsarist permission. The administrative path to approval involved authorization from the St. Petersburg city government, several sections of the MVD, and the Finnish senate if the project involved Finnish territory. These authorities rejected projects for technical, economic, and "formal reasons."[201] The advocates of hydropower did not extend into the ranks of authority.

Although interest in harnessing waterpower for St. Petersburg arose in the late 1880s, the first detailed proposal did not appear until 1894.[202] Basing his plan on Niagara Falls installations, engineer V. F. Dobrotvorskii proposed a Narva River hydroplant to transmit 26 MW 137 kilometers over a 20-kV line, half again as much as the 16 MW from the city's central stations in 1902.[203] His 1896 proposal moved the site to the Imatra River and predicted a lower financial return (6.25 vs. 8 percent).[204] Dobrotvorskii predicted that his project would supply less costly energy than conventional stations. Replacing British coal would save Russia's trade balance 2 billion rubles over eighty years. Finally, Russian engineers would exploit this new technology before foreigners did.[205] Despite these advantages, Dobrotvorskii's proposals remained on paper. Opponents criticized his assessment for technical uncertainties, neglecting the cost of the distribution network, and un-

[201] See "Iz Gazet," *Elektrotekhnicheskoe delo*, 1914, no. 5: 23, for a 1914 Finnish rejection on all three grounds; see TsGIAL f. 1288, op. 9, ed. kh. 195, 3, for a 1910 rejection.

[202] Walther Kirchner, *Die Deutsche Industrie und die Industrialisierung Russlands, 1815–1914* (St. Katharinen: Scripta Mercaturae Verlag, 1986), 98.

[203] V. F. Dobrotvorskii, "Elektricheskaia peredacha energii ot Narvskogo vodopada v Peterburg," *Elektrotekhnicheskii vestnik*, 1894, no. 11: 359–66; T. F. Makarev, "Razvitie oborudovaniia tsentralnykh elektricheskikh stantsii v Peterburge," *Elektrichestvo*, 1912, no. 6: 181; R. R. Tonkov, "Statistika i razvitie elektricheskikh stantsii v S.-Peterburge," *Elektrichestvo*, 1900, nos. 15–16: 217.

[204] V. F. Dobrotvorskii, "Soobrazheniia o peredache elektricheskoi energii ot vodopada na reke Narve v S.-Peterburge," in "Deiatelnost obshchestva," *ZIRTO*, 1894, no. 10: 46–48, and "Snabzhenie g. S.Peterburga elektricheskoi energiei, peredannoi ot vodopadov 'Narvskogo' i 'Imatry'," *Elektrichestvo*, 1896, no. 4: 54–57.

[205] Dobrotvorskii, "Elektricheskaia peredacha," 61.

due reliance on government subsidies and tax exemptions.[206] Four years later, a bitter Dobrotvorskii complained, "This indifference of Russian society is extremely startling to me. Here is St. Petersburg, the center of Russian intellectual life, the flower of the Russian intelligentsia. And suddenly this center is not interested in its economic progress and lets foreign engineers and others look after St. Petersburg's social needs. Even the capital city's self-government was completely apathetic to my project—and who would this project affect more?"[207]

Dobrotvorskii neglected to mention that a hydrostation would adversely affect everyone who benefited from the existing thermal stations and that such risk taking was quite alien to Russian government and business. Nor did he stress that hydrostations cost far more to build than thermal stations, an important consideration for a financially constrained municipality. The engineer continued to promote hydropower, but he directed his attention, equally unsuccessfully, to create laws encompassing waterfalls and long-distance transmission.[208] In 1902, Dobrotvorskii's company and the Finnish company Sitola failed to win the St. Petersburg tram concession.[209] Neither company's proposal benefited from the legal battle, started by Dobrotvorskii, over Sitola's request for MVD permission to construct a transmission line from the Finnish border to St. Petersburg.[210]

Concurrent with and not unrelated to the 1886 Company's growing interest in peat stations for Moscow, major utilities and foreign investors became interested in hydropower for St. Petersburg, which prompted the first efforts by economic elites to establish a legal framework. The 1886 Company and other foreign and domestic firms formed Russian companies to obtain hydroelectric concessions. The major foreign entrant was Imatra, chartered in Brussels with 30 million Belgian francs in 1912. This powerful international syndicate consisted of nine European and Russian banks (one-third of the holdings) and eight electrotechnical companies (two-thirds), including all three St. Petersburg utilities.[211] Imatra was the first firm backed by the foreign finance and technology necessary to construct a commercial hydrostation in Russia.

---

[206] Dobrotvorskii, "Soobrazheniia," 48; Diakin, *Germanskie kapitaly*, 45.

[207] "Khronika," *Elektrotekhnik*, 1897–98, no. 5: 338.

[208] TsGIAL f. 115, op. 1, ed. kh. 14 and 28.

[209] "Elektrotekhnika v Rossii," *Elektrotekhnicheskii vestnik*, 1902, no. 22: 526; TsGIAL f. 1287, op. 44, ed. kh. 251, 23–24, 79–87.

[210] TsGIAL f. 1287, op. 44, ed. kh. 2–10, 60–61, 251.

[211] *Electrical Review*, 29 November 1912, 859; "Khronika gorodskikh upravlenii v Rossii," *Izvestiia Moskovskoi gorodskoi dumy*, 1912, no. 12: 22.

Reversing its previous inaction, the 1886 Company actively aided Imatra's efforts to obtain a hydroelectric concession. The firm's investment of 8 million francs made it the largest shareholder. This activity was one of several moves to guarantee the company's survival. In 1913, the 1886 Company formed the Company for Electrical Regional Stations as a vehicle to expand outside St. Petersburg, prepare for hydropower, and protect some assets from municipal takeover. Discussions with the Petersburg district zemstvo in 1912 for a thirty-year concession aided the decision to restructure.[212] The 1886 Company also started construction of the 20-MW peat-fueled Utkina Zavod station, transferred to the Company for Regional Electric Stations, but the advent of World War I stopped construction.[213]

Imatra represented the German interests of Siemens and Halske, AEG, and their banks. In contrast, French banks supported the competition, Pavel I. Rattner's Peterburgskoe obshchestvo peredachi sily vodopadov (St. Petersburg Company for the Transmission of Power from Waterfalls), which sought a concession for the Vuoks River. The local favorite, Rattner's firm tried to capitalize on nationalist sentiment.[214] A Finnish firm, Fors, submitted another proposal for the Vuoks in 1914, but the Finnish government decided in 1915 to wait until the war's end to decide.[215] The British firm Vickers, a recent arrival in the Russian market, also showed interest.[216]

Predictably, government action lagged behind industry petitions. The Ministry of Transportation was the department most interested in hydropower.[217] Its administration for internal waterways drew up multiple-use projects to improve navigation and harness the hydroelectric potential of the Volkhov, West Dvina, and Dniepr rivers.[218]

---

[212] TsGIAL f. 23, op. 28, d. 2562, 1, 2, 8, 12; M. Giterman, "Elektrichestvo i munitsipalitety," *Izvestiia Moskovskoi gorodskoi dumy*, 1914, no. 9: 77–78; "Iz gazet," *Elektrotekhnicheskoe delo*, 1912–13, no. 6: 22.

[213] "Raionnaia elektricheskaia stantsiia 'Krasnyi Oktiabr'," *Elektrifikatsiia*, 1923, no. 3: 30; A. A. Kotomin and M. D. Kamenetskii, "Obzor deiatelnosti Leningradskogo obedineniia gosudarstvennikh elektricheskikh stantsii 'Elektrotok' za period 1917–1927 gg.," in "Izvestiia Elektrotoka," *Elektrichestvo*, 1928, nos. 1–2: 4.

[214] TsGIAL f. 1288, op. 9, ed. kh. 105, 13–17; N. K., "Imatra v rukakh Belgiitsev" and "Interesy S.-Peterburga i belgiiskaia 'Imatra'," *Peterburgskie vedomosti*, 31 October and 14 November 1912.

[215] TsGIAL f. 1288, op. 9, ed. kh. 105, 2, 45.

[216] Clive Trebilcock, "British Armaments and European Industrialization, 1890–1914," *Economic History Review* 26 (May 1973): 267.

[217] Including power for railroads; see M. A. Tokarskii, "Ispolozovanie padeniia reki Msty dlia elektricheskoi tiagi na Nakolaevskoi zheleznoi doroge," *ZIRTO*, 1902, no. 6: 413–23.

[218] V. D. Nikolskii, "Gidroelektricheskie ustanovki v Shvetsii i Norvegii," *Elek-

The ministry's Volkhov projects of 1902–3, 1909–11, and 1914 formed the basis for the hydrostation finally started in 1918 under engineer Genrikh O. Graftio, a participant in these previous projects. The ministry tried unsuccessfully to resolve the right-of-way issue for transmission lines in 1902.[219] It tried again early in 1913, prodded by new petitions for hydrostations. A few months later, the ministry unsuccessfully asked the Council of Ministers for immediate permission to begin its Volkhov project.[220] The reason for the government inaction lay in the ultimately irreconcilable differences among the Ministry of Transportation, MVD, and MTP about land and waterfalls ownership, compensation for estranged land, transmission lines crossing different governmental jurisdictions, inspection, and ministry oversight.[221] Despite the increasing involvement of powerful economic interests, a law did not pass until the time of the 1917 provisional government. Even then, the debates were not resolved but merely passed to a special committee.[222]

Once again, the state's failure to resolve ministerial differences served the country poorly. Because hydrostations could not proceed without government approval, Russia entered World War I with its capital city dependent on imported coal. The inability of the tsarist government to forge the legal foundation for hydropower, besides retarding the spread of electrification, had grave wartime consequences.

Politics and Visions of the Future

THE USE OF TECHNOLOGIES to promote social and political goals is not new. The development of a new technology often inspires thinking about changes as well as reinforcement of the status quo.[223] Electrification was no exception; like the airplane, its promise and potential

---

trichestvo, 1913, no. 1: 17; V. V. Tchikoff, "Russian Water-Power Possibilities at Dnieper River Rapids," *Engineering News-Record*, 29 May 1919, 1065; *Stroitel pervykh gidroelektrostantsii v St. Peterburge akademik Genrikh Osipovich Graftio, 1869–1949* (Moscow: Izdatelstvo Akademii nauk, 1953), 6.

[219] "Elektrotekhnika v Rossii, " *Elektrotekhnicheskii vestnik*, 1902, no. 19: 450.

[220] "Iz gazet," *Elektrotekhnicheskoe delo*, 1914, no. 4: 21; 6: 19.

[221] Ibid., 1914, no. 4: 21, no. 6: 20; 1915, no. 3: 23; 1916, no. 5: 20–21; 1917, no. 1: 16.

[222] Ibid., 1917, no. 7: 16.

[223] E. g., William J. Lawless, *Microcomputers and Their Applications for Developing Countries* (Boulder: Westview, 1986).

stirred minds and hearts everywhere.[224] Two types of political activities grew from electrification. One was the politics of electric power, which dealt with the establishment and operation of utilities. This normal politics was similar to the politics surrounding a military base or a factory, where the concerns are conventional benefits like profit and prestige. Understanding the motives and means of the various actors does not require a radical shift of thinking. A second was the politics *for* electric power, which envisioned electrification as a tool for basic social, economic, and political change. Such visions are common with new technologies, particularly after their transformatory potential becomes evident but before institutional and societal inertia incorporates them into the existing structure.[225] Often breathtaking in scope and simplistic in depth, visions of technological utopia tend to suffer from the fallacies of total revolution, social continuity, and the technological fix.[226] Although prerevolutionary visions of electrification suffered from all three fallacies, by 1920 the greatly altered political and economic environment had reduced the opposition to wideranging technological changes.

In Russia, the decade after the 1905–6 revolution saw growing efforts by engineers and others to use electric energy "as a mighty factor in contemporary social-economic development."[227] As in Germany, these advocates promoted specific lines of technological development to realize an imagined future.[228] Foremost among them was assistance to small-scale industries to compete with big businesses, the "democratization" of lighting, extension of electric light and power to rural areas, and municipal control of utilities, especially trams. Antiforeign and pro-Russian industrialization elements often intertwined in these proposals, which formed part of a larger technocratic movement

---

[224] Joseph Corn, *Winged Gospel: America's Romance with Aviation, 1900–50* (New York: Oxford University Press, 1983).

[225] Dorothy Nelkin, "Controversy as a Political Challenge," in Barry Barnes and David Edge, eds., *Science in Context: Readings in the Sociology of Science* (Cambridge: MIT Press, 1982), 276–81.

[226] Joseph J. Corn, "Epilogue," in Joseph J. Corn, ed., *Imagining Tomorrow: History, Technology, and the American Future* (Cambridge: MIT Press, 1986), 219–22.

[227] M. Giterman, "Elektrichestvo i munitsipalitety," *Izvestiia Moskovskoi gorodskoi dumy,* 1914, no. 8: 71.

[228] Ulrich Wengenroth, "Die Diskussion der gesellschaftspolitischen Bedeutung des Elektromotors um die Jahrhundertwende," in *Energie in der Geschichte: Zur Aktualität der Technikgeschichte, 11th Symposium of the International Cooperation in History of Technology Committee (ICOHTEC)* (Düsseldorf: ICOHTEC, 1984), 304–11; Edmund N. Todd, "Industry, State, and Electrical Technology in the Ruhr circa 1900," *Osiris* 5 (1989): 243–59.

among Russian engineers and a progressive movement in munici-palities.[229]

The underlying assumption was that the wide and rapid transfer and diffusion of electric light and power would benefit previously un-reached sections of society. Like the 1886 Company, advocates viewed the regional station as the key to the future. Increasing the range of electric stations from tens to hundreds of kilometers would greatly expand the potential base of users. Unlike the utility, these men reacted not to economic threats but to social hopes. By 1914, their proposals had grown into schemes for widespread, regional electrification. Although radical, these ideas were not revolutionary; they incorporated concessions and other elements of the existing po-litical and economic status quo. The proposals reflected their environ-ment; later plans grew more revolutionary and grandiose with the times.

Advocates redefined consumers to include the vast mass of ex-cluded people—workers and others who could not afford lighting from a utility, peasants who lived outside the range of a station. Equally important, just as electricity powered large factories, so its reach would extend to small enterprises. Significantly, and despite the pro-Russian and antiforeign elements of some proposals, most of these ideas and their supporting data originated in the West.[230] Rus-sian engineers knew the work of Georg Klingenberg, AEG's head of powerplant design and pioneer of the regional station in Germany.[231] Russian efforts at municipalization paralleled Western clean govern-ment drives that coupled local control with efficient operations.[232] Russian and Western proposals were somewhat ahead of what was economically and technically feasible, but had a major influence in the wartime and postwar debates over electrification.

One of the most devoted Russian advocates of electric light and power for local use was Perelman. From 1906 through the early 1920s,

---

[229] Rieber, *Merchants and Entrepreneurs*, 355.

[230] E. g., M. Giterman, "Elektrichestvo i munitsipalitety," *Izvestiia Moskovskoi gorodskoi dumy*, 1914, no. 8: 52–71; no. 9: 63–78; no. 10: 54–72.

[231] G. Klingenberg, "Rukovodiashchaia soobrazheniia pri postroike bolshikh parov-ykh elektricheskikh stantsii," *Elektrichestvo*, 1913, no. 3: 99–103; no. 7: 210–14; no. 9: 268, "Generatory i oblastnye tsentralnye stantsii," *Elektrichestvo*, 1914, no. 5: 151–55, and *Large Electric Power Stations: Their Design and Construction* (London: Crosby Lockwood and Son, 1916).

[232] Armstrong and Nelles, *Monopoly's Moment*, 141–58; Anthony Sutcliffe, *Towards the Planned City: Germany, Britain, the United States and France, 1780–1914* (New York: St. Martin's, 1981).

this electrical engineer propagated the doctrine of centrally generated inexpensive electricity for decentralized, small-scale industrial users. In 1906, he proposed a major shift in the energy cycle. Instead of transporting coal by railroad for users, a large powerplant would burn the coal at the mine and transmit the electricity to users over long-distance transmission lines.[233] Technical and economic feasibility were his key determinants: could high-voltage transmission lines provide energy more inexpensively than railroads? In 1906, the answer was negative. A decade later, the VI Section, critiquing the government's five-year plan for railroads, decided otherwise.

Perelman considered that electricity was not, as many city governments and utilities thought, a expensive form of light and power, but the most rational form of energy—if used properly. Proper use meant a monopoly on production, the development of local industries to use electric motors, and balanced loads for low tariffs.[234] Other engineers expanded on Perelman's theses. Strengthening local *kustar* (domestic handicraft) and *remeslo* (artisan handicraft) industries was a prominent clarion call. Electric motors—less costly than their gas and steam counterparts, convenient, clean, and compact—would allow local industries to compete against big, often foreign industrial firms.[235] Two decades earlier in Germany, similar goals had pushed Rudolph Diesel to invent his engine.[236] This was not the "technology with a human face" proposed by E. F. Schumacher over half a century later but a political effort to link Russia's state-driven industrialization with a greater societal distribution of economic and political power.

The drive to spread electric power geographically and economically intertwined with city-based progressive efforts to expand electric light socially from a luxury for the few to a necessity for the many. The technical goals were the extension of the cable network to poorer urban areas and a tariff sufficiently low for workers to afford electric lighting. Interest in this "democratic application" of electricity increased after the introduction of incandescent lights with tungsten filaments in 1908 drastically altered lighting economics.[237] The new

[233] I. Ia. Perelman, *Elektricheskaia energiia i melkoe proizvodstvo* (Moscow: Obshchestvo vzaimopomoshchi russkikh tekhnikov, 1906), 25–26.

[234] Ibid., 38–42.

[235] "Otchet o deiatelnosti IRTO v 1909 godu," *ZIRTO*, 1910, no. 10, 183–84; Giterman, "Elektrichestvo i munitsipalitety," no. 10: 63.

[236] Thomas, *Diesel*, 38–67.

[237] Flekkel, "'Ekonomicheskie' lampochki i populiarizatsiia elektrichestva," 237; "Khronika gorodskikh upravlenii v Rossii," *Izvestiia Moskovskoi gorodskoi dumy*, 1911, no. 8: 50–52.

lamps, which consumed a fourth to a half the electricity of the carbon filament lamp, immediately reduced demand and utility revenue.[238] The new bulbs momentarily stopped Moscow's rapid growth of consumption in 1908.[239] Utilities responded to the loss of income by charging maximum tariffs and increasing the candlepower of the new lamps.[240]

Only after a few years did most utilities lower lighting rates and thereby expand their pool of users.[241] The impetus to lower rates came from such engineers as D. M. Maizel, who predicted beneficial social and political consequences from the "democratization of electric lighting."[242] The key was lower tariffs, which would allow previously marginal customers to afford electric lighting. Maizel, other engineers, and city administrators had to convince utilities to switch from a tariff system that produced high profits per unit of electricity to a system with lower profits per unit but greater output.[243]

Two different philosophies of service produced three distinct views of the urban market. Advocates of low-cost lighting wanted to eliminate or restrict profits to create a large consumer base. Concessions preferred the existing system, which produced large gains from a small base of consumers. For utilities, the issue was not simply lowering tariffs but finding the capital and receiving state authorization to expand. Municipal viewpoints split between those of city treasurers who saw utilities as a means to fill city coffers and service-oriented, progressive politicians and electrical engineers who wanted either low tariffs to expand the consumer base or high tariffs to pay for network expansion and not other city services.

Municipal operations grew in popularity after 1906 as technical risks diminished and the financial needs of municipalities increased. Before 1900, concessions constituted 80 percent of utilities, but that figure dropped to 55 percent by 1914.[244] Equally important was the

[238] L. Dreier, *Posobie pri proektirovanii po elektrotekhnike* (Moscow: Pechatnoe delo, 1912), 2–4; "Razvitie tekhniki lamp nakalivaniia," *Elektrichestvo*, 1916, nos. 7–8: 131.

[239] Kirpichnikov, "Razvitie Moskovskoi tsentralnoi elektricheskoi stantsii O-va 1886 goda," 86.

[240] E. g., increasing the candlepower from 16 to 25–30; see Flekkel, "'Ekonomicheskie' lampochki," 234–36.

[241] Though not, in Flekkel's opinion, as aggressively as needed; "'Ekonomicheskie' lampochki," 235.

[242] "Deiatelnost obshchestva," *ZIRTO*, 1910, no. 5: 71.

[243] Flekkel, "'Ekonomicheskie' lampochki," 237.

[244] "Spisok elektricheskikh stantsii, o kotorykh v VI-m otdele imeiutsia svedeniia," *Elektrichestvo*, 1915, no. 1: 29–30; "Statisticheskie svedeniia . . . za 1914," *Elektrichestvo*, 1917, nos. 4–6: 55.

growing progressive political drive toward municipal control of city services to enhance the quality of life and earn new revenue for local budgets. In a marked change of policy, these progressives willingly assumed debt to extend city services, including electric light, beyond the central city district.[245] Support for the municipalization of utilities came from anticoncession and antiforeign feeling, best displayed in the pages of the liberal *Gorodskoe Delo* (City Affairs), established in 1908.

For cities, ownership promised sorely needed funds and greater control over an important urban service.[246] For advocates of local government, an end to concessions, under which "the interests of the city and its citizens always suffer," would elevate those interests above the profits of [foreign] investors and banks.[247] Opposition to city ownership came from concessions but also from engineers who feared a lack of municipal business gumption and the diversion of profits to other city services instead of reinvestment in electrification.[248] Overall, electrification advocates considered city governments, although lacking sufficient financial means and entrepreneurial spirit, better than concessions.[249]

Only a few electrical engineers, such as Perelman and V. V. Dmitriev, actually moved beyond words. They had to convince non-engineers of the importance of electrical energy, but also to change the mindset of the engineering community "to think what we can do, to escape the passive mode, and to give ourselves not to electric lighting in the West but in Russia."[250] These men spoke as individuals and not for a utility, technical society, or government. They were outsiders without political influence; not until their ideas had circulated for several years and World War I changed the political and economic environment did the electrotechnical establishment embrace the concept of electrification as a transforming social force. Dmitriev,

[245] Hamm, "Khar'khov's Progressive Duma," 18–19.

[246] "Voprosy gorodskoi zhizni v tekushchei literature," *Izvestiia S.-Peterburgskoi gorodskoi dumy*, 1906, nos. 13–15: 181; "Khronika gorodskikh upravlenii v Rossii," *Izvestiia Moskovskoi gorodskoi dumy*, 1907, no. 5: 67.

[247] O. G. Flekkel, "VI Vserossiiskii elektrotekhnicheskii sezd," *Gorodskoe delo*, 1911, no. 5: 451.

[248] V. Smirnov, "Novaia forma ekspluatatsii gorodskikh predpriiatii: Ni kontsessii, ni munitsipalizatsii," *Elektrichestvo*, 1911, no. 10: 252.

[249] V. V. Dmitriev, *Doklad II-mu sezdu lits okonchivshikh Elektrotekhnicheskii institut Imperatora Aleksandra III-go* (St. Petersburg: M. M. Stasulevich, 1910), 6–7: F. A. Dits, "K voprosu ob organizatsii gorodskikh elektricheskikh stantsii," *Elektrichestvo*, 1918, nos. 9–10: 88–92.

[250] Dmitriev, *Doklad II-mu sezdu*, 8.

a socially oriented entrepreneur, viewed electrification in Russia pessimistically in 1910 because of the "comparatively low level of culture among the Russian urban population, the inadequacy of initiative, the fear of the new, and, mainly, absolutely no government advocacy for electrotechnology. To wait for the penetration of this idea in the consciousness of [city dwellers] where many do not even know the word 'electricity' is very difficult."[251]

Dmitriev called for "cultural" specialists—young electrical engineers—to go live in small towns as propagandists and engineers of electrification. They would provide the technical and administrative expertise to make local utilities a reality.[252] Instead of the "People's Will" debacle of 1874, when thousands of idealistic youth went to rouse the countryside to revolution only to be rejected by the peasantry, dedicated but practical electrical engineers would sacrifice positions in big cities to bring power to the people. Dmitriev's plea may have moved hearts, but not minds. His 1910 proposal to create cooperative user-owned utilities for provincial cities too poor to construct plants and too small to attract concessions remained on paper. His 1913 proposal for an alliance of private capital, city governments, and users to work for rapid electrification fared equally poorly.[253] His strategy was correct, but his timing was a decade too early.

Opposition to foreign concessions proved far more effective in galvanizing domestic reaction and eventually action. Early in 1914, negotiations stalemated between the Moscow provincial zemstvo and the Fain company over a powerplant for its suburban tram concession.[254] In response, E. O. Bukhgeim proposed in a newspaper article the most far-reaching scheme yet for electrification, "a new economic factor of equal significance as good communications and roads."[255] His scheme was one of the first Russian plans to electrify an entire region.[256] Fifteen large stations could provide electric energy at one-third of current rates for European Russia, using local coal, peat, hydropower, and southern oil. As Perelman had proposed eight years earlier, lo-

[251] Ibid., 10.
[252] Ibid.
[253] V. V. Dmitriev, *Potrebitelskie elektricheskie stantsii* (Moscow, 1913).
[254] "Munitsipalnoe obozrenie," *Gorodskoe delo*, 1911, no. 18: 1354–55; E. O. [Bukhgeim], "Znachenie oblastnykh elektricheskikh stantsii i nekotorye zadachi zemstv," *Russkie vedomosti*, 27 April 1914, 9.
[255] Bukhgeim, "Znachenie."
[256] P. V. Avtsyn enunciated similar goals in a 1913 talk, but he concentrated on the disadvantages of the 1886 Company monopoly in Moscow and less on alternatives; see *K voprosu ustroistva.*

cally generated electricity and electric motors would replace imported coal and steam engines.

Bukhgeim opposed concessions, but he was realistic enough to realize that no viable alternative existed; a good contract was the best hope. Cooperatives formed by the concessionaire, local zemstvo, and populace would sell motors and other electrical equipment at low prices and provide technical advice and inexpensive credit to increase demand. Kustar and remeslo industries would benefit along with larger, capitalist industries. Elsewhere, more general images of the future predicted mine-mouth generation of electricity, along with a standard current and voltage. The smoke and dust of coal would vanish from cities, factories and railroads would be electrified, and electricity would heat, cool, and light homes.[257]

Although these visions of an electrified future tantalized electrical engineers, the present lagged far behind the future. As with other desired technological futures, those who predicted a radiant future rarely calculated the cost of the transition.[258] Nor did they create the necessary participatory network of users, regulators, and producers to support regional electrification politically and financially.[259] Not until World War I did the necessary connections among the government, industry, and engineering communities form. Not until the destruction of the old order after November 1917 did the new network succeed in making electrification the state technology.

Regional electrification faced institutional and political obstacles, as Elektroperedacha and hydropower demonstrated, but also economic and technical barriers. A major weakness of the advocates for social change was their failure to address the enormous capital and technical demands. The cost of a regional transmission network was daunting: engineer K. K. Shmidt estimated in 1912 that a network with a 75–100-kilometer radius would cost 66 million rubles and produce a dividend of only 2–3 percent at best.[260] The 1886 Company, the largest utility in Russia, had a basic capital of only 40 million rubles and a dividend three times greater.[261] It is no surprise that established firms

---

[257] "Primenenie elektrichestva v budushchem," *Elektrotekhnicheskoe delo*, 1914, nos. 7–8: 25–26.

[258] For examples, see Joseph J. Corn, ed., *Imagining Tomorrow: History, Technology, and the American Future* (Cambridge: MIT Press, 1986).

[259] Michel Callon, "Society in the Making: The Study of Technology as a Tool for Sociological Analysis," in Bijker, Hughes, and Pinch, eds., *Social Construction of Technological Systems*, 83–102.

[260] K. K. Shmidt, "K voprosu o sooruzhenii v Rossii raionnykh stantsii," *Elektrichestvo*, 1912, no. 2: 51–52.

[261] "Po russkim gorodam," *Elektrichestvo*, 1915, no. 1: 32; "Khronika gorodskikh upravlenii v Rossii," *Izvestiia Moskovskoi gorodskoi dumy*, 1912, no. 12: 22.

with large sunken investments were more hesitant about the new technology than individuals who did not report to stockholders or secure millions of rubles of capital.

Was regional service feasible before 1914? For the advocate of electrification, economic profitability played second fiddle to social goals and technological promise. For the utility, the need existed but the outcome was economically dubious. Financially and technically, such a large investment would demand considerable foreign participation. Considering the new technologies, the estimated low rate of return, and legal and political difficulties, the answer is negative. The grandest regional stations and transmission networks were beyond the state of the art in Russia, though not beyond extrapolative planning. Indeed, their proponents viewed these technologies, taken from the West, essentially as a given. Consortia with the necessary technical and financial support did not form until 1912, and only the backers of Elektroperedacha succeeded in building a station. Even then, only the demands of World War I overcame Elektroperedacha's political problems.

Less sweeping dreams fared only somewhat better. Rate decreases, primarily in Moscow and St. Petersburg, did promote the democratization of electric lighting. The 1886 Company increased its subscribers fourfold from 19,659 in 1908 to 78,035 in 1913.[262] Outside the first tier, however, use of electric energy remained very limited. Many city utilities expanded, as did use of electric motors, but neither the extent nor depth of electrification reached the optimistic goals. Electrification advocates failed partly because they acted as individuals without supporting actor networks among local and state governments, consumers, and the electrical engineering community. The lack of receptivity among government and industrial officials also contributed to this failure. They had their own technologies to promote, which reduced available resources, the possibility of building coalitions, and interest in electrification.

Prewar visions of the future were impractical but not impossible. Developments in technology and industrial demand over the next half-decade would bring these visions much closer to economic and technical feasibility. Of greater importance, as the next chapters show, the years of war and revolution established the political foundations for large-scale electrification.

By almost any indicator—number of utilities, per capita consumption, electrical engineers—a gap existed between Russia and the West, a gap that opened in the 1880s and widened greatly in the

---

[262] "Po russkim gorodam," *Elektrichestvo*, 1915, no. 1: 32.

1900s. Why did Russia electrify so slowly in comparison with the West? One factor was the greater urbanization and industrialization in the West, which created the geographic concentrations of users necessary for profitable electric light, power, and traction.

Demand was one part of the story; creating the supply, the other. The economic infrastructure of Western countries was more developed and conducive to the new technology. Financing, the sine qua non of high technology, proved easier with banks, banking houses, and syndicates that evolved to fund electrotechnology. Chronic inadequate financing constrained Russian electrification, as utilities competed in a poorly capitalized market for funds with railroads and industries, and underfunded cities struggled to provide basic services to their swelling populations. It is difficult to understate the importance of capital—and Russia's inability to provide it—for electrification. Another crucial inhibitor was the conservative mindset of state and municipal officials. Not until the 1897 model agreement provided an administrative framework did municipal governments venture into electrification, and even then ministerial regulation and procedure further hindered diffusion and expansion of utilities.

A more general problem was the lack of knowledge about electricity and a lack of promoters who could forge political, technical, and economic alliances. For modernizers, such as Minister of Finance Sergei Witte, the railroad—not the electric lamp—symbolized progress. The Russian electrical world lacked entrepreneurs such as America's Elihu Thomson and Britain's Sebastian de Ferranti. Russia had its inventors and promoters, but no Russian electrical engineer commanded the attention and resources of the nation, government, or industry. The closest were Vladimir Chikolev, whose military activities overshadowed his civilian work, and Mikhail Shatelen, whose prewar interests lay in professionalizing electrical engineering. And electrification had no champions in the national government, a fatal shortcoming in imperial Russia.

Hughes's model of system evolution does not fit prewar Russian electrification, but the mismatch is all the more interesting. The first phases of invention and development never succeeded despite the creativity and effort of talented individuals. Competition and consolidation rarely occurred despite municipal attempts in the first tier. Instead, foreign technology, supported by foreign capital and organization, set the standards for Russian development as technology transferred from a more developed economy to a weaker. The lower indigenous scientific, technical, and financial base suggests that if a country does not develop a technology and infrastructure early, and if it does not make extraordinary efforts, a high degree of dependence on outside technology and finance will ensue.

A Russian national style evolved from hundreds of small decisions and not from an enunciated set of principles. The major technical lacuna compared with the West were hydropower and long-distance transmission, crucial technologies for large-scale electrification. The inability to adopt was political, not technical. A three-tier hierarchy of utilities developed: the first tier had the latest AC equipment, thanks to the large, increasing load and much foreign ownership. High industrial loading and per capita consumption further distinguished this tier from other Russian utilities. The first tier remained on the technical and administrative frontiers, creating an incentive for engineers to remain in Moscow and St. Petersburg instead of moving out to Simbirsk or Tomsk.

Electrotechnology rippled forth from the St. Petersburg–Moscow foundation to cities and towns, bypassing the rural areas and 85 percent of the population. The second tier had AC and DC stations and included equal numbers of municipal utilities and domestic concessions. Most of the foreign-owned concessions were tram systems. Usage of electricity grew and saturation, especially of DC stations, was not uncommon. Local concessions operated the bottom tier in small urban areas. The 1897 model agreement aided the spread of utilities by providing second-tier cities and third-tier towns with an administrative guide. The VI Section served as consulting engineers for many cities, but tsarist Russia never developed the financier-entrepreneurs and financial infrastructure needed to pull utilities beyond a local horizon.

Utilities grew beyond a local level only in proposals for regional stations. The obstacles were more systemic than technical, as hydropower demonstrated, and quite formidable. The lack of holding companies, regional trusts, and other attempts by utilities to rise above the local level save for the 1886 Company is most striking. Even the 1886 Company did not decide to expand into regional stations; outside pressures, primarily higher fuel costs and potential competition, forced its hand.

Electrification had spread widely in Russia, but it sank deep roots in only a few cities. Yet, despite the high rates they charged, utilities often had trouble keeping up with demand. Their policy of conservative growth or expansion within the constraints of existing equipment offered more immediate financial returns than radical growth or geographic expansion. If current demand provided good profits, and the tsarist system of financial and technical approval slowed physical expansion, why should a utility aggressively attempt to expand unless pushed by noneconomic forces? After 1910, such forces, including the concepts of municipality service to its citizens and the democraticization of electric lighting and power, increasingly influenced utilities.

The debate and application of electrification on a larger, more encompassing scale had barely germinated in 1914. World War I and the ensuing revolutions and civil war would ensure that both the challenges and responses would grow greatly in magnitude and consequence. Electrification in Russia had expanded over three decades, but for some dreamers the future promised even more.

CHAPTER 4

The Rise of Electrification,

1914–1917

WORLD WAR I was the single most important factor in the transition from electrification in Russia to Russian electrification. The war drastically worsened the environment for utilities, which lost their technology, financing, and fuel just as military requirements sharply increased demand for electricity. This inability to satisfy wartime needs brought electric power to the attention of state officials and industrialists more effectively than a score of prewar petitions. The war forced the government to recognize the economic importance of electrotechnology, but the state's response was too little, too late, and too disorganized to forge an accommodation with the private sector and electrical engineers.

The rise in the economic importance of electric power stimulated a parallel political rise of Russian electrical engineers. The war hastened the respectability and seeming inevitability of planning Russia's economic future among government officials, industrialists, and engineers. The leaders of the electrical engineering profession, such as Piotr S. Osadchii, proved no exception. Their postwar expectations, framed within the existing political and economic system, laid the groundwork for far-reaching proposals to transform radically the Russian economy by electrification. By war's end, the regional station had started to acquire political as well as economic significance.

Government and Quasi-Government Responses

WORLD WAR I marked a watershed in government–society relations by forcing greater and more intrusive state control of the economy.[1] This increasing role began from necessity as a war expected to last a few months stretched into years, but increasingly it derived from a ideological construct that equated state planning and control with a country's basic interests.

The initial tsarist actions were a response to shortages and not the product of an activist policy. The government's first attempt to improve the industrial contribution to the war was the establishment of the Special Council for the Reinforcement of the Artillery Supply for the Army in May 1915. Four months later, the government created the Special Council for State Defense, with participation by the duma, the state council, ministries, and the private Central War Industries Committee. Direct state control of the economy further increased with the creation of special councils for transport, fuel, and food supply in 1916.[2] Each council signaled the failure of normal procedures to meet the war's unprecedented demands.

As initial enthusiasm subsided into a realization that the war would not end quickly, the scientific and technical communities tried to contribute more than sanitation and food units for the front.[3] As a summer 1915 Imperial Academy of Sciences article enthusiastically proclaimed, "All for war! Victory—a matter for each and every one! Eleven months' knowledge of war has shown that the present war is in significant measure a war of technology and science, a war of the creative forces of the country."[4] Even in the early days of the war, engineers contemplated their potential contributions for the war and postwar period. In October 1914, the prescient mechanical engineer Vasilii I. Grinevetskii, rector of the Moscow Higher Technical School, declared that the two main tasks for engineers were to assist the wartime conversion of industry and to prepare for the postwar period. Grinevetskii accurately predicted the wartime goals of the electrical

[1] William H. McNeill, *The Pursuit of Power* (Chicago: University of Chicago Press, 1982), 317–45.

[2] Peter W. Gatrell, "Russian Heavy Industry and State Defense, 1908–1918" (Ph.D. diss., Cambridge University, 1979), 21, 63, 73; Norman Stone, *The Eastern Front, 1914–17* (London: Hodder and Stoughton, 1975), 204–6.

[3] "Soveshchanie chlenov VI (elektro.) otdela I.R.T.O.," *Elektrichestvo*, 1915, no. 7: 119.

[4] Spravochnoe buro Akademii nauk, *Mobilizatsiia tekhnicheskikh sil*, no. 1: 6, no date but probably printed in August 1915 and published with *Elektrichestvo* and other technical journals. The Library of Congress 1915 volume of *Elektrichestvo* contains three issues.

engineering leadership. For the war effort, scientific and technical societies needed to collect information on the German role in Russia's economy, compare domestic industry with its foreign competitors, and propose economic policy. The main postwar task would be to overcome dependence on German technology, which would demand a unified technical profession, better technical education, and rationalization of Russian industry.[5]

The first attempt to mobilize the country's scientific and technological forces came from the Imperial Russian Technical Society. In late 1914, the society created an information bureau for industry and state and local governments. Inquiries peaked in July 1915 at fifty a month, then decreased as new organizations and more published information fulfilled these needs.[6] By late 1915, newly created organizations had assumed the task of harnessing science and technology for the war. These organizations ranged from small, institutional committees to the nation-spanning war industries committees. An example of the former was the Committee for Military Technical Assistance, established in July 1915 by the Society of Electrical Engineers. The committee produced artillery shells in the St. Petersburg Electrotechnical Institute, repaired instruments, and trained students.[7] The best known example of scientific mobilization was the Committee for the Study of Natural Productive Forces of Russia (KEPS) of the Imperial Academy of Sciences, which functioned through the early years of Bolshevik rule.[8] These attempts to aid the war effort despite state opposition contributed to the technical intelligentsia's growing desire to become involved in postwar policy.[9]

The engineers' enthusiasm to employ their skills paralleled the business support manifested in the war industries committees. Pushed on a reluctant, at times hostile, national government by Mos-

---

[5] V. I. Grinevetskii, *Tekhniko-obshchestvennye zadachi v sfere promyshlennosti i tekhniki v sviazi s voinoi* (Moscow: Biulleten Politekhnicheskogo obshchestva, 1914), 1, 9–15.

[6] "Priostanovka deiatelnosti," *Trudy Komissii po promyshlennosti v sviazi s voinoi*, 1915, no. 12: 1–2.

[7] "Iz gazet," *Elektrotekhnicheskoe delo*, no. 1915, no. 9: 21–22; A. Zalesskii, "Elektromekhanicheskii otdel Petrogradskogo komiteta voenno-tekhnicheskoi pomoshchi," *Elektrichestvo*, 1918, nos. 3–4: 37–40.

[8] Iuri A. Gladkov, *Voprosy planerovaniia sovetskogo khoziaistva v 1918–1920 gg.* (Moscow: Gospolitizdat, 1951), 296; Alexander Vuicinich, *Science in Russian Culture, 1861–1917* (Stanford: Stanford University Press, 1970), 220–22; Kendall E. Bailes, *Science and Russian Culture in an Age of Revolutions: V. I. Vernadsky and His Scientific School, 1863–1945* (Bloomington: University of Indiana Press, 1990), 138–40.

[9] Kendall E. Bailes, *Technology and Society under Lenin and Stalin: Origins of the Soviet Technical Intelligentsia, 1917–1941* (Princeton: Princeton University Press, 1978), 40–43.

cow industrialists and the Association of Industry and Trade, "the principal representative and defender of large-scale industrial interests,"[10] these committees sought to incorporate small- and medium-scale enterprises into the war effort, rationalize industrial policy, and act as clearinghouses for military orders and allocations of resources. The Central War Industries Committee (TsVPK, Tsentralnyi voenno-promyshlennyi komitet), created in May 1915, oversaw regional and local branches. By January 1916, thirty regional and two hundred local committees had formed. Many—fifty-nine in 1917—managed factories.[11] Hostile to finance capital and big business, to government and bureaucracy, the war industries committees were the pinnacle of middle-class business organization. Politically, they marked an extension of the struggle between the government and the business community and an attempt, according to Lewis Siegelbaum, "to project onto an all-Russian scale a plan for victory in the war and through it the future industrial development of the country."[12] Like many organizations formed in troubled times, the promise and vision of these committees exceeded their accomplishments.

The TsVPK established an electrotechnical section in September 1915.[13] A bureau composed of a president, Petrograd physics professor Aleksandr A. Voronov, and three members (including Bolshevik Leonid B. Krasin[14]) coordinated the section's efforts to supply the military with equipment and to secure power, electrical equipment, and materials for private and governmental enterprises.[15] The section represented the electrical engineering elite of Russia in industry, academia, and the VI Section.[16] It worked closely and amicably with industry, government, and the VI Section to discuss common concerns, collect and disseminate information, and attempt solutions. In 1915, for example, a TsVPK survey found 53 MW of surplus power

[10] Ruth A. Roosa, "Russian Industrialists Look to the Future: Thoughts on Economic Development, 1906–17," in John Shelton Curtis, ed., *Essays in Russian and Soviet History in Honor of Geroid Tanquary Robinson* (Leiden: E. J. Brill, 1963), 199.

[11] Gatrell, "Russian Heavy Industry," 63; Lewis H. Siegelbaum, *The Politics of Industrial Mobilization in Russia, 1914–17* (New York: St. Martin's Press, 1983), 50.

[12] Siegelbaum, *Politics of Industrial Mobilization*, 48.

[13] *Mobilizatsiia tekhnicheskikh sil*, no. 1: 3.

[14] The other two were engineer Evgenii Ia. Shulgin and professor Vladimir F. Mitkevich.

[15] "Obshchii ocherk zadach, organizatsii i deiatelnosti elektrotekhnicheskogo otdela Tsentralnogo voenno-promyshlennogo komiteta v sviazi s polozheniem elektrotekhnicheskoi promyshlennosti v Rossii do voiny," *Elektrichestvo*, 1916, no. 3: 33–41.

[16] The VI Section elected eight of the thirty members; "Deiatelnost VI," *Elektrichestvo*, 1915, no. 20: 171; see also "Deiatelnost VI," *Elektrichestvo*, 1916, nos. 7–8: 136.

at provincial power stations theoretically available.[17] The TsVPK sent this information to evacuated factories and firms interested in building defense factories.

Overall, the war industries committees brought more disorganization than organization to the chaos of Russian economic reality.[18] Compared with other parts of the TsVPK and the regional and local branches, the electrotechnical section rates rather favorably. It contributed toward maintaining the Russian electrotechnical industry under increasingly demanding circumstances. The section's relative success stemmed from not operating factories directly and from the community of interests it represented. More important, it represented the first time the electrical engineering community organized to solve national needs. This would be but the first of several efforts.

Imports

THE IMMEDIATE PROBLEM facing utilities was their severe dependence on foreign equipment.[19] Although domestic firms manufactured half of Russia's electrotechnical needs, some equipment, such as instruments, was not manufactured domestically and other equipment was not produced in sufficient quantities to meet demand.[20] The most advanced equipment, such as the large turbogenerators powering first-tier utilities, was all foreign.

Utilities used three approaches to overcome these deficiencies: covert imports from enemy countries, imports from friendly and neutral countries, and domestic production. The first approach was initially most popular. Importers used Switzerland and Stockholm as transfer points, but the tsarist government's insistence on knowing the nation of manufacture eventually blocked these paths, although the MTP decree of 22 September 1915 prohibiting enemy imports exempted electrical and other equipment.[21] Several factors limited imports from

---

[17] TsGIAL f. 1090, op. 1, d. 89, and "Deiatelnost Elektrotekhnich. otdela TsVPK," *Elektrichestvo*, 1916, nos. 5–6: 114–19; see also, "Khronika," *Elektrichestvo*, 1915, no. 19: 357–58.

[18] See Siegelbaum, *Politics of Industrial Mobilization*, 158.

[19] "Vyvoz iz Avstro-Vengerii produktov elektricheskoi promyshlennosti v 1913 gody," *Elektrichestvo*, 1914, no. 13: 336–37; "Vyvoz iz Germanii produktov elektricheskoi promyshlennosti v 1913 gody," *Elektrichestvo*, 1914, no. 14: 347–49.

[20] "Obshchii ocherk zadach," 34; S. A. Gusev, *Razvitie sovetskoi elektrotekhnicheskoi promyshlennosti* (Moscow: Energiia, 1964), 12.

[21] "Otdela TsVPK," *Izvestiia Tsentralnogo voenno-promyshlennogo komiteta*, 15 September

friendly and neutral counties. Gaining government authorization to purchase foreign goods was difficult. The menace of German submarines further limited imports. Only two ports, Vladivostok and northern Archangel, remained open during the war. Only the trans-Siberian railroad linked the former with European Russia, and the latter lacked a good railroad. Increasing domestic production ultimately received the most attention, but it also failed to meet demand. The TsVPK electrotechnical section encouraged domestic production of key materials, such as sheet iron for transformers, and equipment, such as measuring instruments, but the Russian electrotechnical industry could neither meet this demand nor manufacture the high-technology equipment.[22]

Nationalization

THE LARGE CITIES—Petrograd, Moscow, and Kiev—fulfilled their long-standing goals of directly controlling their utilities under the aegis of a broader popular movement to sequester enemy-owned enterprises.[23] Cities, aided by public hostility against Germany and Germans,[24] justified their efforts as the "quick liberation of the population . . . from subjugation by German enterprises."[25] A self-nationalization had already occurred when Russians replaced German and Austro-Hungarians as workers, foremen, and directors of the 1886 Company and Elektroperedacha. Since Russian engineers had been moving up the management hierarchy, this takeover proved more one of degree than of kind, save for the new, direct role of the tsarist and local governments.[26]

The tsarist government moved fairly quickly against the economic

---

1915, 3; M. O. Kamenetskii, *Robert Eduardovich Klasson* (Moscow: Gosenergoizdat, 1963), 118; Boris E. Nolde, *Russia in the Economic War* (New Haven: Yale University Press, 1928), 32–42.

[22] "Obshchii ocherk zadach," 34.

[23] Elsewhere, Dvinsk and Omsk sequestered electric stations, while Baku's efforts to seize Elektricheskaia Sila failed; "Po russkim gorodam," *Elektrichestvo*, 1914, nos. 17–18: 412; 1916, nos. 19–20: 297.

[24] A Moscow newspaper called on consumers to protest against the 1886 Company by not paying their rates; by May 1915, the company had lost 5.4 million rubles; see Kamenetskii, *Klasson*, 113.

[25] "Deiatelnost Moskovskogo gorodskogo upravleniia," *Izvestiia Moskovskoi gorodskoi dumy*, 1915, no. 4: 53.

[26] TsGANKh f. 9508, op. 1, ed. kh. 14, 3.

activity of enemy citizens. A law of 22 September 1914 prohibited the purchase of property by enemy nationals.[27] Further laws in 1915–16 confiscated or closed enemy-owned enterprises.[28] City actions against electric utilities paralleled these state laws. As early as November 1914, the Moscow duma discussed seizing the 1886 Company and Elektroperedacha,[29] and in March 1915 it formally asked the state to liquidate them.[30] The Kiev duma moved along the same lines against the Kiev Electric Company in early 1915.[31] Petrograd pursued the most aggressive route to acquire and unify its three foreign-owned utilities. The city had established a commission in April 1914 to plan the buyout of its three concessions under the terms of the model agreement.[32] Despite financially based arguments to buy only the 1886 Company and opposition from the firms, the duma voted by a two-to-one margin in April 1915 to purchase all three concessions in 1917–18.[33]

The state moved slower than the cities against the utilities. In July and October 1915, the Council of Ministers approved special governing boards for the 1886 Company, Elektroperedacha, and the Company for Electric Regional Stations, the offshoot of the 1886 Company in Petrograd.[34] The Imatra Company had earlier purchased the Company for Electric Regional Stations in an unsuccessful effort to avoid sequestration.[35] The boards consisted of representatives of ministries and the appropriate city government without any industrial or TsVPK participation, despite a request by the Moscow Society of Factory and

---

[27] Before this law, the military council of the Caucasus command rejected the application of a German citizen, Emilio Tsart, for a power plant concession on grounds of awkwardness; "Po russkim gorodam," *Elektrichestvo*, 1914, no. 16: 395.

[28] Nolde, *Russia in the Economic War*, 74–100.

[29] "Po russkim gorodam," *Elektrichestvo*, 1914, no. 20: 444.

[30] "Deiatelnost Moskovskogo gorodskogo upravleniia," *Izvestiia Moskovskoi gorodskoi dumy*, 1915, no. 4: 53.

[31] "Po russkim gorodam," *Elektrichestvo*, 1915, no. 4: 84.

[32] TsGIAL f. 23, op. 28, d. 1610, 170; "Khronika gorodskikh upravlenii v Rossii," *Izvestiia Moskovskoi gorodskoi dumy*, 1915, no. 4: 64–68; "Po russkim gorodam," *Elektrichestvo*, 1915, no. 10: 187; L. V. Sventorzhetskii, "Proekt obedineniia elektricheskikh tsentralnykh stantsii Petrograda," *Tekhnicheskie izvestiia*, 1918, no. 3: 1.

[33] "Zhurnal zasedanii Petrogradskoi gorodskoi dumy," *Izvestiia S.-Peterburgskoi gorodskoi dumy*, 1915, no. 37: 500; "Khronika gorodskogo dela," *Gorodskoe delo*, 1915, no. 6: 338–39.

[34] TsGIAL f. 23, op. 27, d. 841, 74; f. 23, op. 28, d. 1913, 1–2; f. 23, op. 28, d. 2562, 100–104, 128–38; "Po russkim gorodam," *Elektrichestvo*, 1915, no. 4: 83–84; 1915, no. 17–18: 342; 1916, no. 10: 186. The 1886 Company case served as the basis for the two laws empowering the Council of Ministers to appoint receivers; see Nolde, *Russia in the Economic War*, 91.

[35] TsGIAL f. 23, op. 28, d. 2562, 8.

Mill Owners for representation.[36] By February 1917, the state Special Committee on the Struggle with German Dominance completed a plan to sell the stock of the five largest foreign electrical and electro-technical companies, including the 1886 Company and Elektro-peredacha, on 1 June 1917. Old firms would dissolve and new companies form. Russian, neutral, and friendly nationals could have their stock converted, a diplomatic concession to Belgium and France.[37] Private investors, the Moscow and Petrograd governments, and the treasury would each hold approximately one-third of the stock, which would add 30 million rubles of new capital to the utilities.[38] Here, as elsewhere, the tsarist government moved too slowly.

War and the Utilities

THE WAR PLACED the utilities in a "scissors crisis," trapped between greatly increased demand for electric energy and decreased availability of fuel, equipment, and skilled personnel. These problems worsened over time. The sharp reduction in imported equipment halted expansion and adversely affected existing operations. Although the increase in industrial demand affected the first tier worst, utilities everywhere suffered from shortages and restrictions. By mid-1916, the major problems were the loss of trained staff to conscription, stations with too few defense industries to receive official priority for fuel and materials, and the increasing disarray of the economy.[39] Nonetheless, utility output increased through 1916 in response to the war's vast industrial demands. These immediate needs were satisfied at the expense of the health of the physical plant and the nondefense industrialist, businessman, and private individual.

The first shortages, carbon rods for arc lamps, occurred in late 1914. Three factories had supplied the prewar market. German forces occupied one; serious supply problems plagued the second, located in the war zone; and the third had depended on the first for its soot supply.[40]

---

[36] "Iz gazet," *Elektrotekhnicheskoe delo*, 1915, no. 2: 23.
[37] TsGIAL f. 23, op. 28, d. 2562, 116.
[38] "Khronika," *Elektrichestvo*, 1916, nos. 17–18: 276; 1917, no. 1: 25–26. The manufacturing companies were Siemens and Halske, Siemens-Schukkert, and AEG.
[39] "Deiatelnost Elektrotekhnich. otdela," *Elektrichestvo*, 1916, no. 9: 161; no. 12: 222–26.
[40] "Iz gazet," *Elektrotekhnicheskoe delo*, 1914, no. 12: 18–19.

Cities responded by replacing arc lamps with incandescent bulbs and reducing usage.[41] By late 1916, lighting restrictions overshadowed the shortage of rods.

Fuel shortages dominated utility needs by late 1916 because of the greatly increased industrial demand, a railroad system strained by military demands, and a cessation of coal imports to Petrograd. From 1913 to 1915, Petrograd's coal imports dropped more than 90 percent, from 8.7 to 0.6 million tons, necessitating urgent attempts to deliver Donets coal and Black Sea oil to supply the city's vital defense industry.[42] The city government immediately understood its vulnerability. On 21 August 1914, the uprava established a special commission on fuel, the first of many state and local government efforts.[43]

In March 1915, a coal squeeze prompted more systematic procedures to save electricity. Only the transfer of coal stocks from the hospital and Nikolaevsk railroad averted a shutdown of the tram system. By mid-1915, an MTP committee worked on regulations to reduce electricity use, while the naval ministry proposed urgent measures to minimize all nonproductive electrical uses in the capital. Finally, the governor of the region announced restrictions aimed at preserving fuel and electric energy for defense-related factories and workshops. The regulations banned electric lighting for advertisements and building exteriors near street lighting, limited lighting for doorways, and substituted incandescent for arc lights. The more drastic MTP measures limited hours of operations for commercial establishments and greatly curtailed street lighting.[44]

These measures proved insufficient. On 28 October 1915, the city duma authorized its head, D. I. Demkin, to take the necessary measures to ward off a shutdown by the 1886 Company threatened for 30 October. With this threat in the background, Demkin negotiated with the MVD, MTP, and Ministry of Communications for guaranteed monthly allotments of Donets coal.[45]

[41] E. g., Rostov-on-Don and Tiflis; "Po russkim gorodam," *Elektrichestvo*, 1914, no. 16: 395.
[42] L. K. Ramzin, "The Power Resources of Russia," in W. R. Douglas Shaw, comp. and ed., *Transactions of the First World Power Conference* (London: Percy Lund Humphries, 1924), vol. 1, 1249.
[43] Osobaia komissiia po toplivu, *Otchet po operatsiiam s donetskim i angliiskim (poluchennym cherez Arkhangelsk) uglem i drovamu* (St. Petersburg: Petrogradskaia gorodskaia uprava, 1915).
[44] "Po russkim gorodam," *Elektrichestvo*, 1915, no. 6: 116; no. 16: 319; nos. 17–18: 343.
[45] Ibid., no. 19: 358–59.

In 1916, the capital had 105 stations with 193 MW. One hundred one of these were industrial and had an average capacity of 1 MW, small by the standards of the 1886 Company but still larger than most Russian utilities. The city's four utility stations accounted for nearly half of the 193 MW and 60 percent of the 478 MkWh generated.[46] Until the 1886 Company warned its subscribers of power cuts, Petrograd contained twenty-six idle factory plants with 2 MW.[47] Fear of electricity cuts prompted many factory operators to reopen their closed stations, which operated less efficiently than utilities. The consequent demand for electrical equipment and fuel triggered efforts by the city and the TsVPK electrotechnical section to coax guarantees from the 1886 Company to supply these factories.[48] Overloading worsened. By October 1916, the MTP had a list of 1,500 industrial and private users to cut when the industrial load had to be reduced. Despite the disruptions to the city's industrial life, cuts occurred.[49] In February 1917, the 1886 Company had only a week's supply of coal, while the two other concessions literally operated on a shipment-to-shipment basis.[50]

Severe fuel shortages spared Moscow in 1915 because of its closeness to southern fuel supplies and the peat-fired Elektroperedacha. The standoff between the city duma and Elektroperedacha ended in late 1915 when the duma, under MTP pressure, allowed the 1886 Company to receive power from Elektroperedacha.[51] The Moscow uprava proved equally unwilling to enact MTP conservation measures, prompting the commander of the Moscow military district to order their implementation in autumn 1915.[52]

As early as 1915, shortages prompted other cities to mandate cutbacks during peak periods.[53] Utilities justly feared operating at full capacity without reserves, backup equipment, needed maintenance, and a fully trained staff. Despite these problems, surprisingly few blackouts occurred.[54] A major exception was Baku, where the great

---

[46] I. A. Skavani, "Elektrosnabzhenie Leningrada," *Elektrichestvo*, 1924, no. 4: 176.

[47] "Khronika," *Elektrichestvo*, 1915, no. 19: 357–58; no. 20: 375–76.

[48] "Obshchii ocherk zadach," 39–40.

[49] TsGIAL f. 23, op. 27, d. 123, 6–7.

[50] A. L. Sidorev, ed., *Ekonomicheskoe polozhenie Rossii nakanune Velikoi Oktiabrskoi sotsialisticheskoi revoliutsii: Dokumenty i materialy* (Moscow-Leningrad: Izdatelstvo Akademii nauk, 1957), 2: 21.

[51] "Iz gazet," *Elektrotekhnicheskoe delo*, 1915, no. 10: 19; Kamenetskii, *Klasson*, 105–8; Gleb V. Lipenskii, *Moskovskaia Energeticheskaia* (Moscow: Moskovskii rabochii, 1976), 20–23.

[52] "Po russkim gorodam," *Elektrichestvo*, 1915, no. 19: 359.

[53] E. g., Kiev and Odessa; ibid., 1916, no. 1: 31; no. 4: 92; nos. 19–20: 297.

[54] E. g., Rostov-on-Don; ibid., 1916, nos. 19–20: 297.

pressure to increase oil output caused a three-day blackout in summer 1915. The investigation by professor Mikhail A. Shatelen for the Special Council on Fuel blamed the problems on severe overloading and not the suspected sabotage.[55]

By late 1915, the state and private organizations had formed several groups to mitigate fuel shortages. The Special Council on Fuel established a heating commission in June 1916 to decrease consumption, increase fuel availability, and explore alternative fuels.[56] Two of its six sections dealt with regional electric stations, local fuels, and water power.

In December 1915, the Bureau of the Unified Technical Organizations in Moscow founded a heat committee to work with the Special Council on Fuel. Under the guidance of three professors at the Higher Technical School—Karl V. Kirsh, Vasilii I. Grinevetskii, and Karl A. Krug—the heat committee worked to ease fuel shortages and lay the groundwork for postwar development.[57] The committee became a major center of technocratic thinking and alternative fuels research.[58] Its peat section, for example, worked to increase use of the low-quality fuel in the Central Industrial Region to free higher-quality coal and oil for other industrial centers.[59] Peat and oil partially supplanted Donets coal as primary fuels in the Central Industrial Region.[60] For the Petrograd region, wood became the fuel of choice, but as the fuel situation deteriorated the city even considered coal wastes—as a substitute for wood.[61]

The absence of new equipment to replace existing plants or expand capacity aggravated the utilities' problems. MVD authorization for foreign or domestic generators did not guarantee delivery.[62] The rare

---

[55] TsGIAL f. 23, op. 27, d. 841, 69–70. For the full report and the company's response, see ibid., d. 2554, 55–73.

[56] "Khronika," *Vestnik inzhenerov*, 1916, no. 21: 95–96; "Khronika," *Elektrichestvo*, 1917, nos. 7–8: 128–29.

[57] K. V. Kirsh, "Teplovoi komitet pri Moskovskom upolnomochennom predsedatele osobogo soveshchaniia po toplivu," *Vestnik inzhenerov*, 1916, no. 3: 115.

[58] M. L. Kamenetskii, "Pobedenskaia raionnaia elektricheskaia stantsiia," *Voprosy elektrifikatsii*, 1922, nos. 1–2: 99; Alfred J. Rieber, *Merchants and Entrepreneurs in Imperial Russia* (Chapel Hill: University of North Carolina Press, 1982), 399–400.

[59] Kirsh, "Teplovoi komitet."

[60] K. V. Kirsh, "Mery k rasshireniiu dobychi i polzovaniia mestnymi toplivami," *Vestnik inzhenerov*, 1916, no. 15: 513.

[61] For coal, see "Khronika Ts.v.-pr. komiteta," *Izvestiia TsVPK*, 9 September 1915, 3, and 24 October 1915, 2; for coal wastes, see ibid., 1 January 1917, 3, and "Iz gazet," *Elektrotekhnicheskoe delo*, 1917, no. 2: 15.

[62] Kislovodsk and Rostov-on-Don "Po russkim gorodam," *Elektrichestvo*, 1914, no. 13: 352; 1916, nos. 19–20: 297.

utility selling old equipment did quite well.[63] Concern quickly shifted from expansion to maintenance of existing equipment. By 1916, many utilities simply refused to connect new consumers because of a lack of cables and other equipment. For support they could point to resolutions issued by the Special Council on Fuel in December 1915 and June 1916.[64] Tariffs rose in response to higher costs as early as 1914. By early 1916, more than twenty cities had raised rates and restricted nondefense consumption.[65]

Despite these problems, electricity generation grew (see graph 4.1). In the Central Industrial Region, output by eighteen utilities rose by a third from 1915 to 1916.[66] The transfers of a 5-MW turbine from Moscow and a 10-MW turbine to the 1886 Company from the unfinished Utkina Zavod project of the Company for Electric Regional Stations gave Petrograd greater capacity in 1917 than in 1913.[67] The first-tier utilities registered the biggest gains, increasing their output from 1913 to 1916 by 180–185 percent. Petrograd utilities increased their output from 158 to 289 MkWh, Moscow output grew from 131 to 244 MkWh, and Baku output rose from 110 to 191 MkWh.[68] This output would not be surpassed for a decade.

Regional Stations

IF THE DIFFUSION of mass production was the war's most significant industrial innovation, the utility equivalent was the promotion of large electric stations and interconnections to the exclusion of alternative lines of development, a trend that continued after the war.[69] In the United States, government restrictions on coal purchases and the expanded demand for electricity nearly doubled industrial reliance on

[63] Odessa realized a profit of over 100,000 rubles on the equipment from its first station; ibid., 1916, no. 12: 228.

[64] Ibid., nos. 19–20: 298.

[65] Ibid., no. 4: 92; "Iz gazet," *Elektrotekhnicheskoe delo,* 1916, no. 11: 16.

[66] K. A. Krug, *Elektrifikatsiia Tsentralno-promyshlennogo raiona* (Moscow: Teplovoi komitet, 1918), 32.

[67] A. A. Kotomin, "O remonte turbin, proizvedennom na petrogradskikh elektrostantsiiakh," *Elektrichestvo,* 1923, nos. 7–8: 358.

[68] "Biulleten," *Elektrichestvo,* 1930, no. 3: 159. This jump in generation between 1913 and 1916 should be remembered because Soviet statistics use 1913 as the base year for comparing post-revolution generation.

[69] McNeill, *Pursuit of Power,* 330; Thomas P. Hughes, *Networks of Power* (Baltimore: Johns Hopkins University Press, 1983), 285–93; Leslie Hannah, *Electricity before Nationalization* (Baltimore: Johns Hopkins University Press, 1979), 53–62.

Graph 4.1. Electricity generation, 1913–16

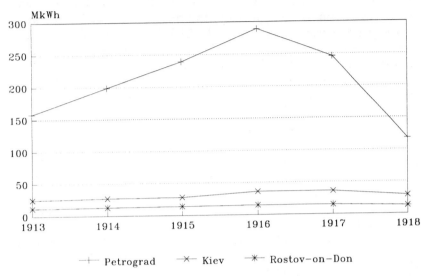

— Petrograd   —×— Kiev   —✳— Rostov-on-Don

Sources: A. A. Kotomin, "Deiatelnost Leningradskikh obedinennykh gosudarstvennykh elektrostantsii (Elektrotok)," *Elektrichestvo*, 1925, no. 5: 329; B. I. Domanskii, "Piat let ekspluatatsii Kievskikh elektricheskikh predpriiatii (1920–1924 gg.)," *Elektrichestvo*, 1925, no. 9: 523; DONGES, "Elektricheskie stantsii Rostova i Nakhichevani v 1922–23," *Elektrichestvo*, 1924, no. 5: 277.

central stations from 1914 to 1919.[70] In Germany, the government financed the construction of a 128-MW plant fueled by brown coal (lignite) to supply the electricity needed for munitions manufacture.[71]

In Russia, the prewar interest in regional stations gathered wartime momentum from fuel shortages, increasing prices, and rapidly growing demand. The concentration of defense industries in Petrograd and Moscow, a reversal of prewar trends toward a more geograph-

[70] C. O. Ruggles, "Some Economic Aspects of the Light and Power Industry," in Arthur H. Cole, A. L. Dunham, and N. S. B. Gras, eds., *Facts and Factors in Economic History: Articles by Former Students of Edwin Francis Gay* (Cambridge: Harvard University Press, 1932), 496–98; Harold L. Platt, "City Lights: The Electrification of the Chicago Region, 1880–1930," in Joel A. Tarr and Gabriel Dupuy, eds., *Technology and the Rise of the Networked City in Europe and America* (Philadelphia: Temple University Press, 1988), 270.

[71] G. Klingenberg, *Large Electric Power Stations: Their Design and Construction* (London: Crosby Lockwood, 1916); Hughes, *Networks of Power*, 287–88.

ically dispersed industrialization, further contributed to this drive.[72]
By 1917, regional stations were planned for Petrograd, Moscow, and
the Donets basin. Most significant, the circle of prospective operators
had expanded from utilities to industrialists and the War Ministry.
The seizure of Moscow and Petrograd utilities and the wartime industrial demands had passed the political initiative to this new set of
economically and politically powerful actors, who proposed building
Russia's largest stations. Electrical engineers now had the ear of industry and parts of government.

As fuel shortages grew, interest in hydropower for Petrograd increased. Governmental bodies, semigovernmental bodies, the city
government, and private firms launched separate efforts in 1916 to
build hydrostations, efforts that ultimately fell victim to bureaucratic
confusion and the worsening economy. The Moscow-based heat committee called for a permanent body to develop a national plan for
"white coal." The fuel commission of the Special Council on Fuel created a section to substitute hydropower for coal. A Petrograd duma
commission proposed a gigantic 200-MW station near Vyborg, Finland, a hydrostation on the Volkhov River, and reconstruction and
unification of the city utilities to meet the expected 1917–20 demand.[73]

By February 1917, two ambitious proposals promised hydroelectric
operations in 1918. The War Ministry planned to harness the Imatra
Falls exclusively for the city's defense industries.[74] The ministry intended to manage the project by a special committee with powers
such as those observed in England and the United States.[75] This proposal marked the first government plan to operate a large hydrostation. If implemented, the project would have reintroduced the military as a key factor in electric power. The second scheme, by the
Finnish firm Fors, intended to use the Vallinkosk Falls to supply both
Petrograd and the Finnish state railroad.[76] A third effort, by Vodopad
(Waterfall), a joint venture by the Petrograd Company for Electric
Transmission and Fors, unsuccessfully sought a concession for the
Imatra and Volkhov rivers.[77]

[72] Martin C. Spechler, *The Regional Concentration of Industry in Imperial Russia, 1854–1917* (Jerusalem: Hebrew University of Jerusalem, 1971), 5, 9.
[73] "Khronika," *Elektrichestvo*, 1916, nos. 17–18: 276; 1917, nos. 7–8: 128–29; "Iz gazet," *Elektricheskoe delo*, 1916, no. 9: 20.
[74] "Khronika," *Izvestiia TsVPK*, 16 February 1917, 3.
[75] "Po russkim gorodam," *Elektrichestvo*, 1917, no. 1: 26.
[76] Ibid.; also "Khronika," *Elektrichestvo*, 1916, nos. 19–20: 296.
[77] TsGIAL f. 634, op. 1, ed. kh. 261, 3–8; "Iz gazet," *Elektricheskoe delo*, 1915, no. 10: 17.

Farther south, large-scale power meant peat and mine-mouth coal plants. The Moscow Society of Factory and Workshop Owners again tried to build a regional station to alleviate a serious situation of inadequate and expensive fuel. The society's peat-burning 25–30-MW plant would provide power at half the prewar cost of Donets coal. The site of the proposed plant later housed the Soviet Shatura station. As in Petrograd, the lack of legal authorization for transmission lines stymied these ambitious plans.[78]

The Donets basin was another emerging area of consumption as coal mining converted from steam to electric power. Generation of electricity had grown tenfold in the decade before 1914, and wartime demand accelerated that trend.[79] In 1916, two companies thought the aggregate demand large enough for regional stations to serve the mines and surrounding cities such as Ekaterinoslav. Ugletok (Coal Current) proposed coal slag, a mining byproduct, to fuel a 20-MW station with potential to expand to 60 MW. The company intended to build two additional stations to free high-quality coal for use elsewhere.[80] The initial plant would be the fifth largest in Russia. The company planned to start operations in January 1918 and received recognition as an enterprise serving the state defense. The Electric Company for the Donets Basin also intended to serve the area's extractive industries and had enlisted some of Russia's leading electrical engineers, including Semen D. Gefter, Aleksandr G. Kogan, and professor Mikhail K. Polivanov.[81]

Despite these industrial efforts, continuing disagreements between the MVD and MTP prevented passage of a law regulating hydropower and transmission lines. The ministries produced conflicting proposals in 1915–16 for transmission lines, and the MTP also developed a broader proposal for an MTP monopoly on long-distance transmission. The MVD criticized this proposal as favoring state over municipal governments and monopoly concessions over municipal operations, as well as for creating undue centralization.[82] Even the

[78] TsGIAL f. 23, op. 27, d. 841, 18, 81–82; "Elektro-kooperativ," *Izvestiia Obshchestva zavodnikov i fabrikantov moskovskogo promyshlennogo raiona*, 1916, no. 8: 13–18; "Khronika," *Vestnik inzhenerov* 2 (15 December 1916): 818.

[79] "Khronika," *Elektricheskoe delo*, 1916, nos. 6–7: 22.

[80] Ibid., no. 12: 18–20; "Khronika," *Elektrichestvo*, 1917, no. 1: 26.

[81] Ibid., 1916, nos. 15–16: 260; nos. 17–18: 276; E. Ia. Shulgin, "Pamiati S. D. Gefter," *Elektrichestvo*, 1928, nos. 7–8: 122; F. Ia. Nesteruk, ed., *Energeticheskaia, atomnaia, transportnaia i aviatsionnaia tekhnika: Kosmonavtika* (Moscow: Izdatelstvo Akademii nauk, 1969), 14–15.

[82] TsGIAL f. 23, op. 27, d. 841, 57–61, 89.

demands of war could not force the MVD and MTP to work together. Military demands focused attention sufficiently for the Council of Ministers in 1916 to create a special conciliation committee to determine MTP–MVD responsibilities for hydropower and long-distance transmission.[83] Only after the February revolution did the committee accomplish its task. In the interrum, Petrograd, Moscow, and other cities suffered.

New Needs, New Dreams

DURING THE WAR, the electrical engineering leadership embraced the idea of widespread electrification and proposed far-reaching state plans for economic and social transformation. The evolution of the electrical engineering leadership into electrification advocates paralleled the growing conviction among industrialists, government officials, and engineers that state planning offered a favorable, directed environment for industry to grow, prosper, and enrich the nation.

In a major shift from prewar MTP interest in economic regularity and predictability, wartime thinking focused on industrial growth in an internationally competitive economy.[84] Competing visions of the future of Russian industry quickly surfaced. Some, such as Vladimir I. Kovalevskii, director of the Department of Trade and Industry from 1892 to 1900 and president of the Imperial Russian Technical Society from 1906 to 1916, almost welcomed the war as an opportunity to throw off the shackles of German economic oppression and develop into an economic power capable of standing up to foreign capitalism.[85] Kovalevskii proposed an immediate state policy to defend and promote all industry, not just large companies. An independent economy demanded the development of Russia's abundant natural resources and the accompanying processing industries. Kovalevskii proposed fourteen measures, including vastly increased credit, easier formation of companies, more technical education, a modernized legal framework, and "quickly harnessing the *energy of falling water*" to

[83] "Khronika," *Elektrichestvo*, 1916, nos. 19–20: 297; 1917, no. 1: 25.

[84] Gregory Guroff, "The State and Industry in Russian Economic Thought, 1909–1914" (Ph.D. diss., Princeton University, 1970), 204.

[85] V. I. Kovalevskii, "Osnovnye nuzhdy russkoi promyshlennosti," *Trudy Kommissii po promyshlennosti v sviazi s voinoi*, 1915, no. 5: 7–8. The opportunity to replace German industry also attracted attention abroad; see, e. g., Ludwig W. Schmidt, "Electrical Development in Russia," *Electrical World*, 26 June 1915, 1720–21.

meet urban and industrial needs.[86] Kovalevskii assumed the continuance of the existing political structure, but he evoked a vision of a radically reformed, rationalized, and supportive political economy to benefit imperial Russia. The Association of Industry and Trade explicitly linked economic growth with national power in its 1915 report on industrial development.[87]

In a 1916 book published by his employer, the Ministry of Finance, Mikhail I. Bogolepov, an economist and chairman of the Committee for the Study of Natural Productive Forces' industrial geography department, publicized the inevitability of large-scale postwar economic reform and thus the need to begin planned development of the country's productive forces. "Economic policy for newly arising processes in the national economy," wrote Bogolepov, "will play the role of midwife and for developed [processes] it will play the role of a smart gardener" to increase and distribute the national wealth more equally. Bogolepov also advocated a decisive revamping of industrial laws, including those hindering hydropower, and decried the change-resistant nature of the Russian polity.[88]

These examples show the growing interest in extending state control over the postwar national economy. The schemes shared common assumptions: that major restructuring of society was needed for Russia to modernize; that the appropriate unit of analysis was not a ministry's purview or an industry but the entire national economy. An unexpressed assumption was that groups currently outside the government—small industrialists, businessmen, engineers, educators, and others—would play a major role in these postwar activities. This goal of greater participation in a more representative government was one reason for the popularity of the war industries committees. These expectations, framed within the existing political and economic system, lay the groundwork for far-reaching, specialized proposals, as in the field of electrification.

A major impetus for this wave of planning came from technical specialists, whose interest lay not in overt political power but in the

[86] Kovalevskii, "Osnovnye nuzhdy russkoi promyshlennosti," 9–12.

[87] *Doklad Soveta o merakh k razvitiiu proizvoditelnykh sil Rossii* (Petrograd, 1915). For the association's wartime activities, see Ruth Amanda Roosa, "Russian Industrialists during World War I: The Interaction of Economics and Politics," in Gregory Guroff and Fred V. Carstensen, eds., *Entrepreneurship in Imperial Russia and the Soviet Union* (Princeton: Princeton University Press, 1983), 159–88.

[88] M. I. Bogolepov, *Puti budushchego: K voprosu ob ekonomicheskom plane* (Petrograd: Izdatelstvo Ministerstva finansov, 1916), 3, 14, 39, 46–52, 57.

rational organization of the economy by experts.[89] The engineer P. Gurevich spoke for many when he declared, "Two years of real war have shown the full necessity of reconstructing the entire Russian national economy on new rational beginnings. *Organization and construction* are the slogans of our time, and by them we will possess the future."[90]

The electrical engineers proved as rational as any. Their wartime proposals shared a common base with their prewar counterparts: regional stations, powered by local fuels and hydropower, would power newly electrified factories and transform Russian industry. Politically, these proposals assumed a transfer of prestige and resources from railroads and heavy industries to electrotechnology and other modern industries. What distinguished the wartime proposals was their greater scale and scope, support from the electrical engineering establishment, and greater promotion. Instead of focusing on increasing capacity, some electrical engineers applied a broader, more systematic and comprehensive approach that envisioned electrification as the foundation of a new, modern Russian economy. For example, the TsVPK electrotechnical section concluded that the rational siting of industry demanded future factories be built not in Petrograd or Moscow but closer to their sources of fuel and materials.[91] This concept had gained interest since the 1910–11 iron shortage. The tsarist railroad plan for 1917–22, discussed below, also advocated establishing industries closer to their raw materials.[92]

The regional station remained the heart of this new thinking, but the rationale and need had expanded. E. O. Bukhgeim revised his electrification proposal in 1915. This extremely significant proposal[93] popularized the economic and "general state significance for all Russia of the organization of electric stations directed to the wide electrification of each given region." The use of local fuels would free the country from dependence on foreign fuels and help the balance of trade.[94] At a conference on Moscow brown coal and peat sponsored by the Special Council on Fuel in November 1915, Gleb M. Krzhizhanovskii, the commercial director of Elektroperedacha, described

[89] Rieber, *Merchants and Entrepreneurs*, 398–99.

[90] P. Gurevich, "Osnovnye voprosy elektricheskoi politiki v poslevoennuiu epokhu v Rossii (okonchanie)," *Elektrichestvo*, 1917, nos. 2–3: 36.

[91] "Otdely TsVPK," *Izvestiia TsVPK*, 1915, no. 6: 2.

[92] TsGANKh f. 3429, op. 1, ed. kh. 1953, 51–51b.

[93] According to G. M. Krzhizhanovskii, *Myslitel i revolutsioner* (Moscow: Gospolitizdat, 1971), 9.

[94] E. O. Bukhgeim, *K ekonomicheskomu osvobozhdeniiu Rossii putem elektrifikatsii ikh territorii* (Moscow: S. P. Iakovlev, 1915), iv, 10, 17, 27.

how, using contemporary American criteria, eight peat-fired stations could supply the entire Central Industrial Region.[95]

The most comprehensive view of a postwar electrified future was offered by Gurevich in *Elektrichestvo* in early 1917. Drawing heavily on Western data, the Swiss-based engineer declared that electricity would replace gas and kerosene for lighting because electric energy demanded "the least expenditure of time, money, and power." As long as a chronic coal shortage hindered the metallurgy industry, he claimed, burning coal for electricity "is completely intolerable from the point of view of the national economy." Consequently, "the Russian electrical industry in the most inflexible development of events will be compelled within ten years and possibly earlier to transfer to a system of large connected regional stations which work on water power or low-quality brown and sulfuric coal and that will be universally accepted as the only rational and advantageous way for generating electric energy."[96]

Using Western activities as examples and for legitimation, Gurevich advocated that state regional stations decrease operating costs for low tariffs and thus high usage. Existing concessions and municipal operations would voluntarily sell the state-produced power since it would be significantly less expensive than generating their own. Gurevich rejected the more radical possibility of state control of production and distribution because of the enormous capital needed to buy out existing utilities and doubts that the central government could market electricity efficiently. To further use of electricity outside the big cities, local and regional governing bodies would handle marketing.[97]

Individuals proposed these early ideas, but in 1916 the VI Section advanced its own postwar scheme. This proposal began with a pro forma request from the IRTO VIII (railroad) Section to review the government's five-year railroad plan for 1917–22, which called for a significant allocation of state resources to expand existing lines and connect new economic regions to the rest of the country.[98] Transporting coal would consume much of the increased capacity.[99] The VI Section offered a technical critique and then proposed the alternate path of

---

[95] G. M. Krzhizhanovskii, "Oblastnye elektricheskie stantsii na torfe i ikh znachenie dlia tsentralnogo promyshlennogo raiona," in *Izbrannoe* (Moscow: Gospolitizdat, 1957), 16.

[96] P. Gurevich, "Osnovnye voprosy elektricheskoi politiki v poslevoennuiu epokhu v Rossii," *Elektrichestvo*, 1917, no. 1: 13–14.

[97] Gurevich, "Osnovye voprosy elektricheskoi politiki (okonchanie)," 34–35, 37.

[98] TsGANKh f. 5208, op. 1, ed. kh. 1, 79–79b/109–110.

[99] *The Russian Government's Plan of Future Railroad Construction* (New York: Youroveta Home and Foreign Trade Company, 1916), 3, 19–21.

electrification to achieve the "comprehensive and most rational development of all the productive forces of the country." The foundation of this development was the "creation of *powerful central electric stations in regions* with rich reserves of fuel or sources of water power to change fundamentally the manner of Russian industry."[100]

The railroad plan assumed the continued transfer of southern coal for northern energy needs, an assumption that the electrical engineers considered inefficient and irrational. At a VI Section meeting on the railroad plan, Evgenii Ia. Shulgin, a longtime proponent of increasing the societal role of the VI Section and electric power,[101] declared that the section's prime consideration was "the use of *all* natural resources of the country and the creation, by the construction of large electric transmission networks, of conditions for industry to use the most direct and economical sources of fuel. . . . This issue, undoubtedly, is no less important than strengthening the output and transportation of fuel."[102]

In conclusion, Piotr S. Osadchii, president of the VI Section and a professor at St. Petersburg Polytechnic Institute, stated that the railroad plan should develop into a larger plan integrating all energy resources and the railroads to meet national needs. According to a paraphrasing of his October 1916 speech, Osadchii said,

> Bearing in mind that one of the decisive factors of this hypothetical plan of railroad construction is the concern about the security of transport of mineral fuel to industrial regions and separate cultural [*kulturnye*] centers as sources of power . . . the correct decision to this question conceivably will be found only after the full study of the possibility of the wide use of electrical transport of mechanical energy from the places where their sources are found—deposits of coal, peat and the so-called "white coal" (water sources)—to the place of consumption.[103]

Instead of railroads hauling coal vast distances for industrial consumption, large mine-mouth electric stations and long-distance transmission would provide secure, reliable energy to users. The new fuels, peat and "white coal," would link with a nation-spanning transmission network as the high technology of electricity replaced the low technology of railroads to power Russia's industries. The VI

[100] "Deiatelnost VI," *Elektrichestvo*, 1916, nos. 17–18: 272.
[101] M. A. Shatelen, "Aleksandr Grigorevich Kogan," *Elektrichestvo*, 1929, nos. 21–22: 592.
[102] "Deiatelnost VI," *Elektrichestvo*, 1916, nos. 17–18: 272.
[103] Ibid., 274.

Section sent Osadchii's proposal to the VIII Section for incorporation into the IRTO report.[104] The country's leading electrical engineering society had just endorsed large-scale electrification as the goal not just for electrotechnology but for the country's economic future.

In a telling demonstration of the importance it now gave electrification, the tsarist government by 1917 effectively controlled the major utilities directly by sequestration and other utilities indirectly by fuel allotments, supply priorities, and consumption restrictions. If the tsarist government had lasted, its control would have increased even more: in early February, the Council of Ministers approved an MTP proposal to regulate the generation and distribution of electrical energy and to sequester enterprises.[105] This measure would have given the government direct operational control over the utilities, a drastic increase of interest and authority since 1914.

The environment for electrification in 1917 differed greatly from that of 1914. Industry and government were more aware of the importance of electric energy and more concerned to remedy its deficiencies. Regional electrification had evolved from the ideas of a few engineers to proposals by the VI Section and industrialists in Moscow and other industrial regions. For the first time, Russian engineers ran the largest utilities and, through the VI Section and new semiofficial groups, promoted electrification as an answer to national problems.

These changes resulted from the war and its ramifications: the expulsion of German managers, greatly increased demand, chronic shortages of oil and coal, the growing unreliability of the train network, and the realization that the existing networks of power were woefully inadequate. Forerunners to these problems existed by 1914, but the war focused attention and pushed the change in the conception of utilities from separate, independent firms to vital elements in the national economy.

The forcing factors were negative, demanding substitution for resources no longer available. Russian utilities met the immediate challenges: electrical generation reached record levels in 1916 despite increasingly adverse conditions. The hidden costs included poorly trained workers, overworked equipment, and growing shortages of materials, parts, and fuels. Equally important, the existing structure of utilities was crumbling under its burden. The cities with the largest industrial loads felt the most pressure, but all utilities suffered.

---

[104] Ibid., 275.
[105] "Khronika," *Promyshlennost i torgovlia*, 11 February 1917, 148.

Before the war, the city was the unit of analysis for electrification. The geographic unit expanded slowly with the construction of Elektroperedacha and proposals for full-fledged regional utilities. During the war, these regional plans expanded, as did the proposed postwar role of electricity. The large size of the planned stations, based on assumptions of wide-scale industrial modernization, broke dramatically from the small stations that constituted the overwhelming majority of Russian utilities. This was an elite vision propagated by electrical engineers to move beyond the reality of electrification in Russia to Western dreams.

In 1916, the VI Section formally accepted and promoted the concept of regional stations fueled by low-quality fuels to transform industry and relieve the transportation system. Thus the electrical engineering leadership in Russia had already enunciated and promoted the basic themes of Soviet electrification four years earlier. Although revolutionary compared with prewar thoughts and tsarist postwar plans, these grand visions of an electricity-based future were accompanied by a confident sense of inevitability. In 1917, Gurevich thought state electrification inevitable within a decade and possibly earlier.[106] The fall of the tsarist government and two revolutions made him seem a pessimist. Electrification was the official policy of the state only three years after he wrote.

---

[106] Gurevich, "Osnovye voprosy elektricheskoi politiki (okonchanie)," 36.

CHAPTER 5

Feasting Eyes, Hungry
Stomachs, 1917–1920

DURING 1917–20, Russian society underwent massive upheavals un-
paralleled since the Time of Troubles three centuries earlier. It was an
era of revolution, of terror, of starvation, of epidemics, and of that
harshest of conflicts, civil war. In February 1917, the tsarist government disintegrated and a duma-based provisional government ruled
until the Bolsheviks seized control in November. From 1918 to 1920,
civil war raged, sharpened by foreign intervention and a trade em-
bargo, until ended by the ruthless autarkic mobilization of war communism. Economically, the country deteriorated from bad to worse.
Paralleling industry, electricity production dropped sharply in 1918–
20 and did not regain prerevolutionary levels until the mid-1920s.
Only the extraordinary efforts and creativity of utility workers kept
electricity flowing.

Yet these years were also a time of bold visions and utopian dreams
that sharply contrasted with the economic and social devastation of
the half-deserted cities and hunger-wracked countryside. Planted in
1914–16, the seeds of state electrification now blossomed as the loci of
decision making and control shifted from the city and utility to a receptive national government. Electrical engineers created a network
of state agencies for electrification and spread the gospel of electrification for societal transformation. By 1920, electrical engineers could
claim the establishment of theoretical rationales and organizational
frameworks for large-scale electrification, the basis of a political alliance with Lenin and other leaders, and state approval for four regional stations. That visions outdistanced reality and initial plans
were unrealistic should not be viewed too harshly: at times, civil war

threatened Petrograd and Moscow, hardly a situation conducive to long-term planning. For the utilities to survive was an impressive technological achievement; the advancement of national electrification was no less an accomplishment in the political realm.

The Provisional Government

THE EIGHT MONTHS of the provisional government proved a heady and frustrating time for utilities and electrical engineers. Taking advantage of their vastly expanded freedom, electrotechnical organizations flourished inside and outside the government as part of a broader movement of technical specialists into important and often new government positions in a demonstration of the forces supporting the February revolution.[1] The provisional government as well as its rival, the Petrograd City Soviet, continued the wartime increase of state control by advocating a centrally planned economy and establishing the Central Economic Committee "for the coordinated implementation by individual departments and institutes of all measures regulating the economic life of the country."[2] As in other areas, the Provisional Government fell short.

The government considered the utilities of Petrograd and Moscow too important to leave to municipal control. In the confused days after the tsar abdicated, the state duma's executive committee gave the TsVPK electrotechnical section responsibility for the "correct and uninterrupted work" of the Petrograd utilities.[3] In the Central Industrial Region, the MTP acted as general overseer.[4] One function of these new authorities was to work with other government bodies to restrict consumption as fuel supplies worsened. In spring, the military command and the Moscow uprava agreed to cut consumption by a quarter to a half of 1916 levels.[5] In September, the provisional government imposed severe restrictions on Petrograd and Kiev users, including industry, and required the use of metal filament bulbs instead of the less efficient coal filaments.[6]

---

[1] Alfred J. Rieber, *Merchants and Entrepreneurs in Imperial Russia* (Chapel Hill: University of North Carolina Press, 1982), 406.

[2] Robert Paul Browder and Alexander F. Kerensky, eds., *The Russian Provisional Government: Documents*, 3 vols. (Stanford: Stanford University Press, 1961), vol. 2: 676; Leon Smolinski, "Planning without Theory," *Survey*, July 1967, 108–10.

[3] "Otdely Tsentralnogo v.-pr. k-ta.," *Izvestiia TsVPK*, 13 March 1917, 4.

[4] "Iz gazet," *Elektrotekhnicheskoe delo*, 1917, no. 2: 20.

[5] "Po russkim gorodam," *Elektrichestvo*, 1917, nos. 7–8: 129–30.

[6] "Khronika," *Elektrichestvo*, 1917, nos. 13–14: 178.

The April creation of the MTP Council for Electrotechnical Affairs (SED, Soveshchanie po elektricheskim delam) gave the electrical engineering community its first government foothold. The 22-member SED, headed by VI section president Piotr Osadchii, represented the major electrical organizations and important outside groups.[7] Like many initiatives of the provisional and Soviet governments, the origins of the SED lay in the old regime—in this case, an unsuccessful attempt in the last months of tsarism by the MTP and the VI Section to centralize all state electrotechnical activities in the high-level Section for Electricity (Otdel Elektrichestva). The reluctance of other ministries to surrender their authority resulted in the compromise creation of the SED to "work out exact measures in electrotechnology" and serve as a "base of activities for the future Section for Electricity."[8]

The SED's immediate priorities were to create a legal framework for regional stations and to establish the Section for Electricity. It focused on regional stations to "alleviate the current fuel and transportation crisis which will continue until the restoration of peace" and because "the further development of Russian industry must be linked with its electrification."[9] The rationale for the section was to realize the industrial potential of substituting mechanical for human power, a change possible only by electrification. The West had demonstrated the increasing economic importance of electrification and regional stations; Russia must follow. The MTP section should manage all electrical oversight, including MVD functions, for optimum efficiency. An interdepartmental committee of experts and representatives of concerned organizations, based on French experience, would direct the section.[10]

Like prerevolutionary groups, the SED failed to establish a single, central government organization for electrification, as would other committees in the early 1920s. Although at times a fanatic confounding of organization with implementation, the attempts to seize the commanding bureaucratic heights were a logical response to a government in which one ministry could bury a rival ministry's proposals in an interdepartmental morass. In a society so heavily dominated by the tsarist—and later Soviet—state, advocates assumed that progress toward the establishment of supportive high-level governmental bodies meant progress toward electrification.

[7] Including the VI Section, the TsVPK electrotechnical section, the Free Economic Society, zemstvo and city unions, and the Association of Trade and Industry, the last represented by Leonid Krasin; TsGIAL f. 23, op. 27, d. 70, 51.
[8] "Ot redaktsii," *Elektrichestvo*, 1917, nos. 9–10: 131–32.
[9] Ibid., 1917, nos. 11–12: 147–48.
[10] TsGIAL f. 23, op. 27, d. 70, 24–26, 34–35.

New organizations also formed outside the government. The first congress of electric utilities, suggested in 1915, met in May 1917 and formed the Soiuz elektricheskikh stantsii obshchestvennogo polzovaniia (Union of Electric Stations for General Use).[11] Modeled after the German Vereinigung Deutscher Elektrizitatswerke (Union of German Electric Utilities), which contained twenty Russian utilities in 1914, the union spoke for management.[12] Its top priority was to alleviate the utilities' worsening financial situation, with lesser interests in the growing problems of fuel, materials, and labor.[13]

In March and April, electrical engineers from the VI Section established the Union of Electrotechnicians to participate in "the great tasks of national economic reconstruction and growth [and] raising the productivity of national labor."[14] Based on the German Verband Deutscher Elektrotechniker and the American Institute of Electrical Engineers, the union sought to unite all sectors of electrotechnology. Osadchii was elected president, ensuring that the electrical engineering community literally did speak with one voice.[15]

The SED kept extremely busy with new tariffs, taxes, labor questions, and a myriad of other issues.[16] An immediate utility concern, pushed by the Union of Electric Stations, was raising tariffs to cover the increased costs of labor and fuel.[17] The SED also worked on an electrical lighting tax, which set rates by type of use and, most interesting, efficiency.[18] Unlike previous attempts, this tax met with no major objections, reflecting both different circumstances and initiators.

The major legal initiative of the provisional government was a land-

[11] A. Vulf, "Chto mogut predpriiat nashi elektricheskie stantsii dlia oslableniia posledstvii voiny i naloga na elektricheskoe osveshchenie?" *Gorodskoe delo*, 1915, no. 6: 312. This union was also known as the Soiuz tsentralnykh elektricheskikh stantsii (Union of Central Electric Stations).

[12] "Khronika," *Elektrichestvo*, 1917, nos. 11–12: 162.

[13] "Ot redaktsii," and "Khronika," *Elektrichestvo*, 1917, nos. 9–10: 131–32, 145.

[14] "Ot redaktsii," *Elektrichestvo*, 1917, nos. 7–8: 111.

[15] "Deiatelnost Soiuza elektrotekhnikov," *Elektrichestvo*, 1918, nos. 3–4: 40–45.

[16] See, e. g., the agenda for the 13 September meeting: TsGIAL f. 23, op. 15, d. 641, 1–2.

[17] "Polozhenie ob izmenenii uslovii dogovorov na otpusk energii elektricheskimi stantsiiami obshchestvennogo polzovaniia," *Elektrichestvo*, 1917, nos. 11–12: 160. The order was "labor and fuel"; ibid., 147–48. Before the February revolution, published complaints focused on the rising costs of fuel and materials and the difficulty of finding skilled workers. The high cost of labor rose to prominence only after February. In an insert in the October *Elektrichestvo*, the editor stated that costs had increased ninefold from August 1914 to the February revolution but sixfold in the previous eight months.

[18] "Proekt polozheniia o naloge na elektricheskuiu energiiu," *Elektrichestvo*, 1917, nos. 15–16: 192–93.

mark law authorizing hydropower and long-distance transmission. In July, the SED published a temporary statute for public discussion, twenty-three years after the first hydroelectric proposal for St. Petersburg.[19] The statute gave the government ultimate command of all water sources with corresponding rights of estrangement and occupation. An MTP committee for the use of waterfalls would direct development, with some local participation.

Although hailing the statute as a significant advance, critics complained that it failed to protect local governments and to distinguish sufficiently between private and public interests.[20] The MVD charged that local governments would lose control of the "significant profits" from municipal stations that supported other municipal activities.[21] These criticisms delayed the statute's approval. A September meeting of the VI Section and Union of Electrotechnicians with the Society for the Study of the City Economy (Obshchestvo izucheniia gorodskogo khoziaistva) conceded the MVD criticism and the need for more exact legal work, but it approved giving the MTP "the authority for industrial projects of state significance," which included the "extremely urgent" and "unobstructed normal development in Russia of networks for electric transmission."[22] The provisional government approved the temporary statute, but the government's fall precluded action.[23] Actual implementation did not begin until an August 1919 Soviet statute established administrative responsibilities.[24]

The provisional government lasted only eight months but laid the institutional groundwork for future state electrification. The formation of the SED, the first high-level government body for electrification, reflected both the rising power of the electrical engineering community and a larger technocratic mindset. Although a sharp jump in support from the tsarist government, the provisional government's actions in many ways continued, not broke with, the trend toward electrification as a state technology. The early years of Soviet rule accelerated this trend as advocates established organizational niches and convinced others of electrification's importance to socialism.

[19] "Vremennoe polozhenie ob ispolzovanii vodnykh sil," *Elektrichestvo*, 1917, nos. 9–10: 143–44.

[20] "Ot redaktsii," *Elektrichestvo*, 1917, nos. 11–12: 147; nos. 15–16: 181–82.

[21] TsGIAL f. 23, op. 27, d. 70, 1.

[22] "Khronika," *Elektrichestvo*, 1917, nos. 15–16: 184.

[23] F. I. Nesteruk, ed., *Energeticheskaia, atomnaia, transportnaia i aviatsionnaia tekhnika: Kosmonavtika* (Moscow: Izdatelstvo Akademii nauk, 1969), 274.

[24] "Polozhenie ob ustroistve i ekspluatatsii Elektroperedach," *Sbornik dekretov, postanovlenii, rasporiazhenii i prikazov po narodnomu khoziaistvu*, 1921, no. 3: 471–74.

The Soviet Government

THE NEW SOVIET GOVERNMENT struggled on many fronts during the 1918–21 period of war communism. A "heady brew of visionary speculation and hardheaded, desperate measures designed to make the economy work and the Soviet regime survive,"[25] according to Richard Stites, war communism was a command economy based on coercion, centralization of production and distribution, mobilization of labor, requisitioning of peasants' products, and the elimination of markets.[26] The malleability of the economy and the power of ad hoc planning through allocating resources demonstrated by War Communism greatly influenced Soviet planning a decade later.[27]

A mix of tsarist, World War I, and Marxist elements went into the bouillabaisse of the new government. Commissariats, governed by the cabinet Council of People's Commissariats (SNK, Sovet narodnykh komissariatov), replaced ministries. The All-Russian Central Executive Committee appointed the council. The Congress of Soviets stood above the SNK, but its infrequent meetings meant that the SNK held the real reins of power. Most relevant to electrification was the December 1917 creation by SNK of the Supreme Council for the National Economy (VSNKh, Vysshii sovet narodnogo khoziaistva), which was charged with organizing the national economy and state finances.[28] Although originally envisioned as guidance and planning bodies, the VSNKh and its regional counterparts became control and administration units under the pressures of the civil war, the revolutionary impetus of war communism, and the example of the wartime special councils.[29]

The VSNKh quickly developed an electrotechnical bureaucracy, the Electrotechnical Section and Elektrostroi within the Committee for State Construction. The Electrotechnical Section (ETO, Elektrotekhnicheskii otdel), formed in December 1917 to handle the production of electrical energy and the underlying manufacturing base,[30] had goals

---

[25] Richard Stites, *Revolutionary Dreams: Utopian Vision and Experimental Life in the Russian Revolution* (New York: Oxford University Press, 1989), 46.

[26] Moshe Lewin, *Political Undercurrents in Soviet Economic Debates* (Princeton: Princeton University Press, 1974), 77–83.

[27] Eugene Zaleski, *Planning for Economic Growth in the Soviet Union, 1918–1932* (Chapel Hill: University of North Carolina Press, 1971), 34.

[28] "God borby," *Narodnoe khoziaistvo*, 1918, no. 11: 1.

[29] M. A. Manuilov, "Skhema organizatsii VSNKh," *Tekhnika*, 1918, no. 1: 17–19; Maurice Dobb, *Soviet Economic Development since 1917* (New York: International Publishers, 1948), 56, 111–13; Zaleski, *Planning for Economic Growth*, 24–27.

[30] "Khronika," *Narodnoe khoziaistvo*, 1918, no. 2: 19. For subsections in 1918, see Man-

that ranged far beyond operating existing utilities. The announcement of the ETO's creation proclaimed that only a network of regional stations could

> place the Russian economy on the level demanded by the international situation. . . . . For the reconstruction of the national economy after the end of the war, the first question is about receiving inexpensive energy by the directed and the planned [construction] of regional electric stations of high voltage (120,000 V) from "white" (waterfalls), "grey" (peat), and black coal. . . . These plans on a state scale already have been worked out (Germany) or are being worked out in all countries.[31]

Three stages of nationalization completed the legal subordination of utilities started in World War I. The new government nationalized Elektroperedacha and the 1886 Company stations in December 1917 and January 1918.[32] The general nationalization of 28 June 1918 covered all utilities with a capital of more than one million rubles.[33] The last stage in 1918–19 extended the reach of the ETO subsection for electric stations, Elektrotok (Electric Current), over state, municipal, concessional, and private stations. This last stage met resistance from the People's Commissariat of Internal Affairs (NKVD, Narodnyi komissariat vnutrennikh del), which represented city and local authorities loathe to lose control over their utilities. In August 1919, the ETO received authority for power stations, ending a year of struggle over local or central control of utilities. For the moment, the centralizers had won.[34]

The ETO did not monopolize state electrification. The VSNKh established the technocratically oriented Committee for State Construction (KGS, Komitet gosudarstvennykh sooruzhenii) in May 1918 to "work out a plan, establish priorities, fulfill and execute state con-

uilov, "Skhema organizatsii VSNKh." For 1919, see "Deiatelnost elektrotekhnicheskogo otdela V.S.N.Kh.," *Narodnoe khoziaistvo*, 1919, nos. 1–2: 41–42.

[31] "Khronika," *Narodnoe khoziaistvo*, 1918, no. 2: 19.

[32] Anatolii V. Venediktov, ed., *Natsionalizatsiia promyshlennosti i organizatsiia sotsialisticheskogo proizvodstva v Petrograde (1917–1920 gg.)* (Leningrad: Izdatelstvo Leningradskogo universiteta, 1958), 1, 114–17; *Narodnoe khoziaistvo*, 1918, no. 5: 63.

[33] "Dannye otdela upravleniia predpriateliami pri kollegii organizatsii proizvodstva VSNKh," *Narodnoe khoziaistvo*, 1918, no. 4: 45–46; "Dekrety i postanovleniia po narodnomu khoziaistvu," *Narodnoe khoziaistvo*, 1918, no. 5: 68.

[34] "Khronika," *Narodnoe khoziaistvo*, 1918, no. 2: 19; "Deiatelnost Prezidiuma VSNKh," *Narodnoe khoziaistvo*, 1919, no. 8: 70; nos. 9–10: 83; 1920, nos. 9–10: 33; Anatolii V. Venediktov, *Organizatsiia gosudarstvennoi promyshlennosti v SSSR* (Leningrad: Izdatelstvo Leningradskogo universtiteta, 1957), 1, 500–501, 524–25; Venediktov, *Natsionalizatsiia promyshlennosti i organizatsiia*, 1, 117–18.

struction, and also survey all projects of state construction and general works treated in VSNKh sections."[35] At the behest of the Moscow-based electrical engineering leadership, one of the eight initial KGS sections handled electrotechnical construction and became Elektrostroi (Electric Construction) in October 1918.[36] This success partly stemmed from the fact that Gleb Krzhizhanovskii served as the first KGS director.[37]

Elektrostroi planning subsections united previously independent prerevolutionary projects, such as the electrification of the Donets basin by the nationalized Electric Company for the Donets Basin.[38] The most prominent part of Elektrostroi, at times eclipsing it, was the Central Electrotechnical Council (TsES, Tsentralnyi elektrotekhnicheskii sovet), formed in October 1918 as "an institute of permanent consultants" using their expertise "for the best and quickest explication of technical and drafting questions about new electrotechnical construction."[39] The TsES was an elite body comprising the same engineers who had formed the SED, the Permanent Committee of the All-Russian Electrotechnical Congresses, and VI Section consulting committees.[40]

Unlike the VI Section committees, the TsES intended to initiate and not just respond. It brought scientists and engineers together to work on defined national needs. Achieving and harnessing this unity was a concern of Soviet leaders, whose efforts and experiments during these years did not always succeed.[41] The TsES gave this cooperation a public prominence, government support, and political power that electrical engineering lacked previously. This prestigious body for elite electrical engineers more than compensated for the effective dis-

[35] "Vremennoe polozhenie o Komitete gosudarstvennykh sooruzhenii V.S.N.Kh. i sostoiashchikh pri nem uchrezhdeniiakh," *Narodnoe khoziaistvo*, 1918, no. 5: 70.

[36] N. P. Bogdanov, "Skvoz grozy i buri," in *Sdelaem Rossiiu elektricheskoi* (Moscow: Gosenergizdat, 1961), 42. Hereafter, *SRE*. See also, Vasilii Iu. Steklov, *Razvitie elektroenergeticheskogo khoziaistva SSSR: Khronologicheskii ukazatel* (Moscow: Energiia, 1970), 17.

[37] "Gleb Maksimilianovich Krzhizhanovskii," *Elektrichestvo*, 1972, no. 2: 2.

[38] TsGANKh f. 3429, op. 1, ed. kh. 1162, 42–44, 48–49; "Skhema VSNKh po Komitetu gosudarstvennykh sooruzhenii," *Tekhnika*, 1918, no. 4: 22; "K istorii elektrifikatsii RSFSR," *Krasnyi arkhiv*, 1939, no. 95: 19.

[39] "Postanovlenie Soveta narodnykh komissarov o TsES," in *K istorii plana elektrifikatsii sovetskoi strany*, ed. Iuri A. Gladkov (Moscow: Gospolitizdat, 1952), 4.

[40] P. S. Osadchii, M. A. Shatelen, and A. G. Kogan, "Tsentralnyi elektrotekhnicheskii sovet za tri goda ego sushchestvovaniia," in *Elektrifikatsiia Rossii: Trudy 8 Vserossiiskogo elektrotekhnicheskogo sezda v Moskve 1–10 oktiabria 1921* (Moscow: Gosizdat, 1921), 1, 134.

[41] See Kendall E. Bailes, *Technology and Society under Lenin and Stalin* (Princeton: Princeton University Press, 1978), and Robert A. Lewis, *Science and Industrialization in the USSR* (London: Macmillan, 1979).

appearance of the VI Section, which ceased publishing *Elektrichestvo* in 1918 due to shortages of paper and power.[42] The ETO and Elektrostroi overlapped functionally, a situation typical of the early years of Soviet power. The appointment of Bolshevik Piotr G. Smidovich, who had worked at the 1886 Company Moscow station, to head both bodies minimized potential conflict. Smidovich also served as a member of the Moscow Soviet of Workers and Peasants Deputies, the new municipal duma. In January 1919, a formal division of responsibility stabilized the framework of state electrotechnical activities. The ETO received the utilities and manufacturing industries, and Elektrostroi received the authority to construct all powerplants for factories and railroads "having general national or regional significance."[43]

The KGS, headed by Krzhizhanovskii and guided by its task of construction, advocated national planning: "It is indispensable to have a state plan even for five years ahead—a plan of development of the productive forces of Russia to answer the questions: what needs to be done earlier, what later, and how to do it?"[44] Fuel and transportation shortages impelled early planning schemes to focus on a region and not the country. Ideally, products would be manufactured and consumed locally to maximize regional self-sufficiency and to minimize interregional transport. The concept of the economically autarkic region grew from necessity, not from an inherent superiority of approach; from existing work on regions, particularly the Central Industrial Region; and from the civil war division of the country.[45]

One indicator of the growing widespread interest in electrification beyond the ETO and Elektrostroi was the appearance of other electrification units at the local, regional, and national levels.[46] The ETO established a group in April 1918 to work with the regional councils

[42] "Ot redaktsii," *Elektrichestvo*, 1918, nos. 1–2: 1. The 1918 issues were hand-corrected and printed on low-quality paper.

[43] "Polozhenie ob upravlenii kupnymi obedineniiami natsionalizirovannykh predpriatii pri VSNKh," *Narodnoe khoziaistvo*, 1919, no. 5: 58; "Deiatelnost elektrotekhnicheskogo otdela VSNKh," *Narodnoe khoziaistvo*, 1919, nos. 1–2: 41; "Postanovlenie SNK o peredache elektrostroitelstva v vedenie KGS VSNKh," 9 May 1919, in Gladkov, ed., *K istorii plana*, 10.

[44] A. E. Makovetskii, "Glavneishaia zadacha gosudarstvennogo stroitelstva v Rossii," *Tekhnika*, 1918, no. 1: 2.

[45] Ibid., 3; N. Charnovskii, "Znachenie raionirovaniia promyshlennosti dlia ekonomicheskogo stroitelstva strany," *Narodnoe khoziastvo*, 1920, nos. 11–12: 17.

[46] Iuri A. Gladkov, *Ocherki stroitelstva sovetskogo planovogo khoziaistva v 1917–1918 gg.* (Moscow: Gospolitizdat, 1950), 223, 424; Gladkov, *Voprosy planerovaniia sovetskogo khoziaistva v 1918–1920 gg.* (Moscow: Gospolitzdat, 1951), 307–8.

for the economy to "unite all work preparation, project planning, and execution of the electrification of industry in the large sense of the word and operation of regional electric stations."[47] Planning bureaus were established for the Dniepr, Donets, Ukraine, Siberia, and Caucasus regions, but the more developed Northern and Central Industrial regions advanced the furthest.[48]

Continuing their prerevolutionary concentration of political and economic power, Moscow and Petrograd dominated regional planning. The Northern Region housed the most active planning bureau, which was established in January 1918 to gather information on existing stations and plan future stations.[49] The Commissariats for Land and Communication also contained electrification bureaus to locate energy resources and determine the requirements of existing and potential customers.[50]

The Fuel Crisis

THE FIRST YEARS of Soviet power saw the creation of an impressive semicoordinated array of electrification agencies. The problems they faced, however, appeared even more formidable. In 1917–21, the fuel crisis, or "fuel hunger," dominated economic life as the railroad network nearly collapsed and White forces occupied Russia's two major energy regions, the Donets coal basin and the Baku oil fields.[51] By 1920, fuel shortages had created a whole series of crises in the economy.[52] Utilities did not escape: output plunged sharply below prewar

[47] "Organizatsiia komitetov po elektrifikatsii ekonomicheskikh raionov," in Gladkov, ed., K istorii plana, 37–39; "Otchet o deiatelnosti otdela elektrotekhnicheskoi promyshlennosti pri VSNKh," in "Khronika," Narodnoe khoziaistvo, 1918, no. 5: 30.

[48] G. O. Graftio, "Pervoocherednye raboty v oblasti elektrifikatsii," 29 June 1918, in Gladkov, ed., K istorii plana, 3; "Doklad nachalnika rabot po shliuzovaniiu dneprovskikh porogov inzhenera V. L. Nikolaia v KGS ot 27 fevralia 1919," Krasnyi arkhiv, 1939, no. 95: 21–23; Gladkov, Voprosy, 297.

[49] M. Grandov, "Ne otstavaite ot Pitra," Bednota, 24 April 1920, 1; Venediktov, Natsionalizatsiia promyshlennosti i organizatsiia, 1: 235–38.

[50] "Spisok VSNKh," Narodnoe khoziaistvo, 1918, nos. 6–7: 75–79; "Doklad nachalnika rabot," 22–23; "Deiatelnost Prezidiuma VSNKh," Narodnoe khoziaistvo, 1920, nos. 1–2: 34.

[51] Ton-miles dropped by a factor of four between 1917 and 1918 and did not recover 1917 levels until 1923; Transport i sviaz SSSR: Statisticheskii sbornik (Moscow: Gosstatizdat, 1957), 32.

[52] A. I. Rykov, "O polozhenii narodnogo khoziaistva," Ekonomicheskaia zhizn, 4 March 1921, 2.

Graph 5.1. Electricity generation in three cities, 1913–21

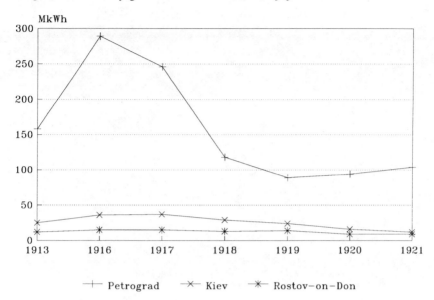

—+— Petrograd   —×— Kiev   —✳— Rostov–on–Don

Sources: A. A. Kotomin, "Deiatelnost Leningradskikh obedinennykh gosudarstven-nykh elektrostantsii (Elektrotok)," *Elektrichestvo*, 1925, no. 5: 329; B. I. Domanskii, "Piat let ekspluatatsii Kievskikh elektricheskikh predpriiatii (1920–1924)," *Elektrichestvo*, 1925, no. 9: 523; DONGES, "Elektricheskie stantsii Rostova i Nakhichevani v 1922–23," *Elektrichestvo*, 1924, no. 5: 277.

levels (see Graph 5.1). In 1919–20, the major utilities came danger-ously close to collapsing. That they did not was a tribute to the engi-neers and technicians who kept powerplants operating, partly by con-verting to local fuels.

Although the new state gave utilities priority for fuel, forage, food, and transportation, shortages eased only slightly.[53] Fuel was only the most obvious problem. In 1920, inoperative transformers remained at Moscow substations because only two of fourteen repair cars worked and inspectors refused to go on foot because they lacked boots. A request for seventy screwdrivers produced one. Utilities needed wood for fuel but also to replace poles torn down for barricades dur-ing the 1917 fight for Moscow and later for fuel. Hunger became a

[53] "K protokolu 1 iiunia 1918 goda," *Narodnoe khoziaistvo*, 1918, nos. 6–7: 34; "De-iatelnost elektrotekhnicheskogo otdela VSNKh," *Narodnoe khoziaistvo*, 1919, nos. 1–2: 42; "Postanovlenie Sovnarkoma o neocherednom snabzhenii elektrostantsii mate-rialami, toplivom, prodovolstviem, furazhom i lichnym sostavom," 3 February 1920, in Gladkov, ed., *K istorii plana*, 250–51.

Extracting peat with sharp implements. Courtesy of the Hoover Institution.

serious problem after mid-1919 when supplying peat to Elektroperedacha depended on feeding the workers. In November 1920, Moscow
stokers were too famished to work a full shift.[54]

The newly established extraordinary commissions for fuel supplies
in Moscow and Petrograd unified the utilities of the two cities into
networks early in 1919. Petrograd's Petrotok had to embark on a project of technical unification before the city's utilities could function co-
operatively.[55] In contrast, the Moscow Unified State Electric Stations
(MOGES, Moskovskoe obedinenie gosudarstvennykh elektricheskikh
stantsii) was a true unified system whose stations operated on the
same current and frequency, allowing the transfer of electric power
for more efficient operations. The monopoly position of the 1886

[54] TsGANKh f. 9508, op. 1, ed. kh. 12, 1–4; TsGANKh f. 3429, op. 1, ed. kh. 1162,
39–40; "Naruzhnoe osveshchenie," *Kommunalnoe khoziaistvo*, 1921, nos. 1–2: 29.

[55] L. V. Sventorzhetskii, "Proekt obedineniia elektricheskikh tsentralnykh stantsii Petrograda," *Tekhnicheskie izvestiia*, 1918, no. 3: 1–7.

Table 5.1. Fuel use in Petrograd utilities, 1917–21

Year	Wood[a]	Fuel oil[b]	Coal[b]	Local coal[b]	Peat[b]
1917	1	151	146	—	—
1918	14	91	65	—	—
1919	567	31	22	2	18
1920	600	20	13	8	19
1921	302	65	5	9	200

Source: A. A. Kotomin, "Vliianie toplivoi politiki na rabotu kotelnykh petrogradskikh elektrostantsii v periode 1917–1921 gg.," *Elektrichestvo,* 1922, no. 1: 16.
[a] Thousand cubic meters.
[b] Thousand tons. Coal is Donets coal; local coal has a lower caloric value.

Company and its prewar attempt to capture the industrial market had produced a citywide cable grid, and Elektroperedacha provided some independence from southern oil. Without its peat station and unified grid, Moscow's power supply would have suffocated.

One sign of desperation in 1918 was the concerted substitution of wood and peat for the coal and oil no longer available. Instead of local fuels supplying regional stations, wood and peat now fired existing central stations in last-ditch efforts to keep some electricity flowing. Five electric stations in Moscow and Petrograd converted to wood in 1918–19; in the second half of 1919, they received 750,000 cubic meters of wood, or 21 percent of the 3,570,000 cubic meters distributed to approximately 400 establishments.[56] Wood provided over half the fuel for Petrograd in 1919–20 until supplanted by oil and peat in 1921 (see Table 5.1). The new government also institutionalized peat as a state priority with the establishment of the Main Peat Committee (Glavnyi torfianoi komitet), a peat academy, survey and construction groups, and a congress of peat workers.[57] This institutionalization benefited primarily regional stations; given a choice, local stations preferred higher-quality fuels.

The now-standard problems of resource allocation and distribution

[56] A. Lomov, "Toplivnyi golod i nasha toplivnaia politika," *Narodnoe khoziaistvo,* 1919, no. 6: 14.
[57] The Main Peat, Coal, and Oil committees fell under the VSNKh fuel section; "Glavnyi torfianoi komitet," *Narodnoe khoziaistvo,* 1919, no. 3: 46–49, 50-53; nos. 11–12: 65. The peat academy emerged from a reorganization of the Main Peat Committee science-student section; "Torfianaia akademiia," *Narodnoe khoziaistvo,* 1920, nos. 3–4: 21–23. The congress met in March 1919; "Sezd torfianikov," *Narodnoe khoziaistvo,* 1919, no. 4: 60–61.

hindered fuel production everywhere.[58] Consequently, the large wartime increases in production of the Central Industrial Region's local
fuels, peat and brown coal, disappeared. Peat production fell by a
third from 1913–16 to 1919. Production of brown coal doubled from
1913 to 1917 before sinking to one-third above prewar levels in 1919.[59]
These shortages devastated the economy. By mid-1918, "the catastrophic state of fuel affairs in the Petrograd region threatens northern industry with an inevitable death."[60] Despite the conversion of
stations to wood and peat, Petrograd power generation decreased
sharply from 289 to 89 MkWh from 1916 to 1919. Petrotok cut some
users off completely and limited others to two or three hours of
power daily in the worst months.[61]

Successfully changing the Petrograd fuel supply demanded significant changes in preparation and burning. The wartime shift from British and German coal to Donets coal and fuel oil required only minor
boiler alterations because the quality of fuel remained high. The civil
war shift to local fuels—wood, peat, and Borovicho coal—demanded
major changes in boilers. The enormous consumption of wood (fifty
wagonloads a day for the six converted boilers of the 1886 Company
station alone) also necessitated a new supply system to deliver the
wood directly to the boilers.[62]

Creating the most efficient boiler for wood required much experimentation. The initial, extremely unsuccessful tests of the Kirsh fire-
box were overshadowed by the development of an efficient wood-
burning boiler by professor Tikhon F. Makarev, a member of the
Moscow heat committee.[63] By enabling utilities to burn wood, the
Makarev boiler kept the Moscow and Petrograd utilities operating

[58] M. Nemenskii, "Doklad po obsledovaniiu i revizii del glavnogo torfianogo komiteta
otdela topliva VSNKh," *Narodnoe khoziaistvo*, 1918, nos. 6–7: 367.

[59] For peat, from 1,580 to 1,080 kilotons; for brown coal, from 324 to 810 and back to
480 kilotons; see M. Progorovskii, "Podmoskovskii kamennougolnyi bassein," *Narodnoe
khoziaistvo*, 1918, no. 5: 10; see also "Dobycha torfa v tsentralno-promyshlennom raione
v techenie 1913–1919 g. (v sezone)," *Narodnoe khoziaistvo*, 1920, nos. 5–6: 11.

[60] "Toplivosnabzhenie i perspektivy petrogradskoi promyshlennosti," *Narodnoe
khoziaistvo*, 1918, no. 5: 47.

[61] A. A. Kotomin and M. D. Kamenetskii, "Obzor deiatelnosti Leningradskogo obedineniia gosudarstvennikh elektricheskikh stantsii 'Elektrotok' za period 1917–1927
gg.," in "Izvestiia Elektrotoka," *Elektrichestvo*, 1928, nos. 1–2: 4.

[62] A. A. Kotomin, "Vliianie toplivoi politiki na rabotu kotelnykh petrogradskikh elektrostantsii v periode 1917–1921 gg.," *Elektrichestvo*, 1922, no. 1: 14–16.

[63] A. A. Kotomin, "O mekhanicheskoi pagache drovianogo topliva v kotlakh bolshoi
paroproizvoditelnosti," *Elektrichestvo*, 1924, no. 11: 538–42.

during the civil war.[64] A less heralded but equally ingenious technical response to crisis was the development of heaters used in lightbulb sockets: "The natural stirring of the population to defend itself from extinction created an unusual blossoming of 'kustar' production of all possible heating instruments" to the point that nearly every apartment in Moscow had one. These heaters kept the population alive and per capita consumption at prewar levels despite the loss of industry and population.[65]

Unlike the Lodygin lamp, which was essentially created in isolation, the Makerev boiler grew from a specific, highly visible state need. Close ties among the Moscow heat committee, utilities, and government fused a defined need, the financial and technical means to solve it, and the scientific and industrial expertise to implement the solution. The kustar heater illustrated that a low-technology approach operating without the resources of government could also satisfy social needs. Both boiler and heater evolved in an environment where specific problems demanded immediate resolution and the technical resources and skilled people existed. These technical improvisations could not, however, overcome the economic inertia of a society in collapse.

During 1919–20, the Extraordinary Commission for Electricity Supply for Moscow desperately sought more fuel.[66] Railroad problems limited oil supplies, and inadequate personnel, food, and extraction equipment caused peat shortages. These shortages forced the 1919 alteration of boilers to burn wood at every station except the factory stations near Elektroperedacha. The problem increasingly was not only insufficient fuel but inefficient utilization of that fuel. Utility efficiency dropped by half due to poor fuel (including green wood), poorly maintained equipment, boilers burning fuels for which they were not designed, and poor working conditions.[67] As elsewhere, the lack of spare parts and maintenance proved major problems.[68]

Output in 1919 dropped to less than 40 percent of the 1916 peak, as

[64] A. A. Glazunov and L. I. Sirotinskii, "Uchastie Moskovskogo energeticheskogo instituta v sozdanii i razvitii elektricheskikh sistem SSSR," *Elektrichestvo*, 1955, no. 11: 13.

[65] V. I. Ianovitskii, "Elektrosnabzhenie Moskvy i blizhaishie perspektivy v etoi oblasti," *Elektrichestvo*, 1922, no. 1: 23.

[66] TsGANKh f. 3429, op. 1, ed. kh. 1162, 38–40, 167, 195.

[67] Ibid., 13–14.

[68] "Moskovskoe kommunalnoe khoziaistvo vo vtoroi polovine 1921 g.," *Kommunalnoe khoziaistvo*, 1922, no. 10: 20.

Graph 5.2. Moscow electric output, 1912–1921

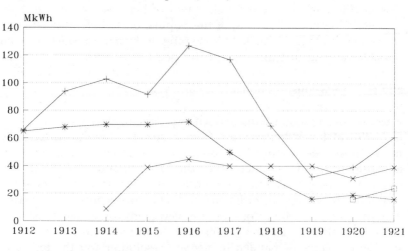

Source: S. P. Stafrin, "Rabota elektricheskikh stantsii Moskovskogo raiona za 1922 g.,"
Elektrichestvo, 1922, no. 1: 36.

Graph 5.2 shows. From 1919 to 1921, trams ceased operating and industrial use dropped sharply.[69] Only Elektroperedacha and nearby
factory stations, hastily connected to the peat plant's 70-kV transmission lines, kept Moscow supplied with electricity.[70] During the first
months of 1919, Moscow utilities had only a few days' supply of oil;
operations literally hinged on a single trainload of oil and very strict
user restrictions.[71] The capital actually suffered a two-day blackout in
December 1919 caused by a lack of oil that closed the 1886 Company
station, bad weather, and administrative mismanagement that sent
peat to the wood-burning tram station. So much peat piled up that no
room existed for wood, closing the tram station. A freeze and snow-

[69] A. Reidel, "Elektrosnabzhenie Moskvy i Moskovskoi gubernii," *Kommunalnyi rabotnik*, 1921, nos. 3–4: 12; nos. 5–6: 12.
[70] "Postanovlenie VSKNh ob upravlenii elektricheskikh stantsii Bogorodskogo raiona," *Sbornik dekretov, postanovlenii, rasporiazhenii i prikazov po narodnomu khoziaistvu,*
1921, no. 3: 119.
[71] Vs. Vasilevskii, "Vtoroi sezd Sovetov narodnogo khoziaistva Severnogo raiona,"
Narodnoe khoziaistvo, 1919, nos. 9–10: 82.

storm prevented access to the wood stockpile. Meanwhile, Elektrope-redacha ran short of peat.[72]

These shortages accelerated interest nationwide in unifying electric stations into grids. Three types of unification emerged. The first unified all the utilities of a city into a single network, like the MOGES. The second type linked city lines via substations to regional stations, as proposed by the VI Section in 1916. The third type, a product of war communism and the need for electricity, connected unused stations at closed factories to supply nearby factories and other users.[73] All three types had Western antecedents, but extreme need and limited resources pushed factory-to-factory unification to the center of Soviet attention.[74]

Local Construction

WHILE SHORTAGES crippled generating capacity and planners worked on large-scale electrification schemes, more than 250 small second- and third-tier stations were built (Table 5.2).[75] This local growth was based significantly on stations started in tsarist times.[76] Like other utilities, these stations, with average capacities of 90 kW for urban and 18 kW for rural stations, suffered from inadequate funding, equipment, supplies, and personnel, as seen in the declining station size over time. Like the VI Section in tsarist times, the newly created ETO and TsES gave technical assistance when requested.[77] This growth occurred almost unnoticed by the electrical engineering establishment, which viewed these small stations as an unchallenging line of technological development. Although they lacked the theoretical justification, technological challenge, and actor network that made regional stations so attractive, the rural stations represented an alternative line

[72] Information from the holdings on Iu. V. Lomonosov in the Leeds Archive, courtesy of A. J. Heywood. See also TsGANKh f. 3429, op. 1, ed. kh. 1162, 195.

[73] In early 1919, Elektrostroi planned this technically "very simple" task for several factories; "Iz doklada upravliaiushchego upravleniia elektrosooruzhenii A. V. Vintera predsedateliu Sovnarkoma V. I. Lenin," 21 April 1919, *Krashyi arkhiv*, 1939, no. 96: 28.

[74] Thomas P. Hughes, *Networks of Power* (Baltimore: Johns Hopkins University Press, 1983), 289–92; Edmund N. Todd, "Industry, State, and Electrical Technology in the Ruhr circa 1900," *Osiris* 5 (1989): 251–57.

[75] Different sources present slightly different data. E. g., E. N. Moiseenko-Velikaia Gorev, "Proizvodstvo elektricheskoi energii," *Elektrifikatsiia*, 1923, no. 8: 6–16.

[76] A. Rykov, "Itogi sovetskogo stroitelstva," *Narodnoe khoziaistvo*, 1920, nos. 17–18: 12.

[77] Gladkov, *Voprosy*, 298.

Table 5.2. New stations, 1917–20

Year	City		Village		Total	
	No.	Average kw	No.	Average kw	No.	Average kw
1917	5	819	5	20	10	320
1918	16	77	12	32	28	58
1919	41	68	20	16	61	51
1920	60	48	103	17	163	28
	122	90	140	18	262	52

Source: V. L. Levi, "Ocherki po elektrosnabzheniiu R.S.F.S.R.," *Elektrichestvo*, 1922, no. 1: 29.

of technological development with quite different political implications. By the early 1920s, these stations had acquired theoretical justification and a political network of supporters who viewed them as a means to transfer electric light and power together with state power rapidly to the countryside.

A sense of the vast distances yet to be covered came in a January 1919 issue of *Ekonomicheskaia Zhizn (Economic Life)*, a VSNKh newspaper.[78] An article described the advantages of electric over kerosene lighting, an argument that had appeared forty years earlier in *Elektrichestvo* with the arrival of incandescent lights in St. Petersburg. The audience had changed but the arguments remained remarkably constant over time. As electric light diffused into new areas, justifications and lures for users preceded it. The article also hailed local projects that represented the first application of electricity for the people in whose name the October revolution was carried out. Despite industrial needs, the first new areas electrified were workers' apartments and nearby streets to demonstrate the benefits of the new government.[79] Such new construction and interconnections indicated a new priority—serving the unserved. These projects represented a step toward the democratization of electric energy, albeit under very different circumstances than its prewar advocates had anticipated.

[78] "Po Rossii," *Ekonomicheskaia zhizn*, 1 January 1919, 6.
[79] E. g., "Gubernskie sezdy zaveduiushchikh otdelami kommunalnogo khoziaistva," *Kommunalnoe khoziaistvo*, 1921, nos. 1–2: 17; on Kharkov, see "Khronika mest," *Kommunalnoe delo*, 1924, no. 10: 68; on Moscow, see K. Lovin, "Kratkii predvaritelnyi otchet o deiatelnosti 'MOGES' za 1923–24 operatsionnyi god," *Elektrichestvo*, 1924, no. 11: 571.

Electrical Engineers and Planning

THE EARLY YEARS of Communist rule expanded the coverage and goals of the wartime discussions about planning and electrification. Two accompanying trends were the further growth and acceptance of a technocratic, rationalist approach to postwar, now post–civil war, reconstruction, and the transfer of the electrical engineering leadership from Petrograd to Moscow.

By October 1917, key elites in government, industry, and the engineering communities had experienced some state planning and control and expected more.[80] The mix of Marxist utopian visions and revolutionary dreams further strengthened interest in planning, as did new journals such as *Ekonomika, Trud, i Tekhnika* (Economics, Labor, and Technology), published by the Moscow Soviet.[81]

Knowledge of foreign activities and ideas, particularly German wartime state planning, stoked this growing Russian interest.[82] A major German influence was Karl Ballod, an economics professor in Berlin. His *Der Zukunftstaat* (The Future State) described how to organize a centrally planned socialist economy. The 1898 German edition appeared in several Russian translations from 1903 to 1906. The second edition, published in Germany in 1919 and in Russia in 1920, helped convince Lenin and others of the feasibility of a planned economy.[83] Ballod's ideas influenced Russians as early as 1898, when Aleksandr I. Ugrimov, later a Soviet specialist on agricultural electrification, heard Ballod lecture at Leipzig.[84] While Ballod influenced Russian concepts of planning, another German, Georg Klingenberg, influenced Russian concepts of regional stations.[85] German experience and theory were

[80] Ruth A. Roosa, "Russian Industrialists and 'State Socialism', 1906–17," *Soviet Studies* 23 (January 1972): 414–16; Bailes, *Technology and Society*, 22–23, 424.

[81] Stites, *Revolutionary Dreams*, 36–37, 45.

[82] Peter Rutland, *The Myth of the Plan: Lessons of Soviet Planning Experience* (London: Hutchinson, 1985), 11; Leon Smolinski, "Planning without Theory," *Survey* (July 1967): 116; Alek G. Cummins, "The Road to NEP, The State Commission for the Electrification of Russia (GOELRO): A Study in Technology, Mobilization and Economic Planning" (Ph.D. diss., University of Maryland, 1988), 52–53.

[83] V. I. Lenin, *Collected Works*, 4th ed., ed. Institute of Marxism-Leninism, 60 vols. (Moscow: Progress Publishers, 1964), 32: 140; E. H. Carr, *The Bolshevik Revolution*, vol. 2 (London: Macmillan, 1952), 373; Smolinski, "Planning without Theory," 117–20; Roger W. Pethybridge, *The Social Prelude to Stalinism* (London: Macmillan, 1974), 43; "Karl Ballod," *Bolshaia sovetskaia entsiklopediia* (Moscow: Partiinoe izdatelstvo, 1926), 4: 539–40.

[84] Aleksandr I. Ugrimov, "Moi put i rabota v GOELRO," *SRE*, 85.

[85] G. M. Krzhizhanovskii, "Tekushchie voprosy elektrifikatsii," *Elektrichestvo*, 1922, no. 2: 4; see also L. Dreier, *Zadachi i razvitie elektrotekhniki* (Moscow: Pechatnoe delo,

important because they gave Russians, Bolshevik and non-Bolshevik alike, the justifying legitimacy of foreign interest as well as guidelines.

A elite group of electrical engineers in Moscow and Petrograd promoted an agenda, based on Western concepts and activities, to develop regional stations and industries based on technically and economically rational criteria that they defined. These electrical engineers were not apolitical, but their politics was the politics of expertism, where they—as the most qualified people—would make the key decisions. The activities of these men fit William M. Evan's definition of technocrats: "The engineer imbued with the technocratic vision believes, on the one hand, in the capacity of technology to solve all social problems without recourse to value considerations, and, on the other hand, in the importance of integrating engineers into the political structure of society."[86] Underlying the faith in planning were two important technocratic concepts: the most rational geographic distribution of industry proximate to resources and transportation, and the careful, concomitant creation of detailed information with which to allocate resources.[87] The best example of the technocratic approach was Vasilii I. Grinevetskii's extremely influential *Poslevoennye perspektivy russkoi promyshlennosti* (Postwar Perspectives on Russian Industry), published in Kharkov in 1919 and reprinted in Moscow in 1922. A Russian pioneer in planning methodology that integrated engineering and economic criteria, Grinevetskii wrote the most comprehensive analysis of nation-wide economic reconstruction.[88] An elaboration of his wartime analysis of engineers' responsibilities, *Poslevoennye perspektivy* reached a larger audience because of the much greater desire for radical reconstruction and the depth and breadth of his information and proposal. Grinevetskii described the country's primary task as, in descending priority, rebuilding supplies of fuel and raw materials; reconstructing and developing transportation; restructuring the technical organization of industry; improving the quality and efficiency of labor; defending the domestic market from foreign imports;

1919), 22, and Boris Kushner, *Revoliutsiia i elektrifikatsiia* (Petrograd: Gosudarstvennoe izdatelstvo, 1920), 13.

[86] William M. Evan, "Engineering," *International Encyclopedia of the Social Sciences* (New York: Macmillan, 1968), 5: 79.

[87] K. A. Krug, *Elektrifikatsiia Tsentralno-promyshlennogo raiona* (Moscow: Teplovoi Komitet, 1918), 3–4.

[88] Leon Smolinski, "Grinevetskii and Soviet Industrialisation," *Survey*, April 1968, 107.

and changing the market's economic structure.[89] To secure fuel and raw materials, he wrote, "in first place should be the *electrification of industry* supplied from *regional stations*, working on local poor-quality, but therefore cheap, fuel."[90] The now-standard troika of peat, Moscow brown coal, and hydropower would power these stations. Unlike other proposals, Grinevetskii's did not neglect financing. Despite the danger of economic and political enslavement, he saw an enormous flow of foreign capital—15–20 billion gold rubles over ten to twelve years—as a "vital condition for the economic revival of Russia and reconstruction of its industry."[91]

Despite political differences, Soviet planners in the 1920s used Grinevetskii's wide-ranging concepts, data, and methodology for three reasons. First, his mesh of engineering and economic rationalism fit well into the technocratic Soviet concept of planning. Second, students and colleagues of Grinevetskii worked for GOELRO and Gosplan, which provided a direct conduit for the transfer of his work. Third, his information was unsurpassed. In 1919–20, members of the heat committee, TsES, the VI Section, and other technocratically oriented engineers, economists, and planners needed reliable information. They realized, in Grinevetskii's words, that planning must be based on *"real data on the economy of the whole country,"* because "future possibilities are made feasible and tightly limited by the past and present."[92] Knowledge, if not power, at least constituted the building block for planning and action.[93] Starting with the publication of utility statistics in 1910 by the VI Section and continuing through the Glavelektro geological expeditions of the 1920s, gathering data was a major function of every electrification and planning agency.

A major contribution to this effort was Karl A. Krug's *Elektrifikatsiia Tsentralno-promyshlennogo raiona* (Electrification of the Central Industrial Region), published in 1918 by the heat committee. Krug, spurred by a fuel crisis that he viewed as chronic and not temporary, outlined a plan of electrification noteworthy not only for its assumptions about the potential of inexpensive electricity but also for the rich array of

---

[89] V. I. Grinevetskii, *Poslevoennye perspektivy russkoi promyshlennosti*, 2d ed. (Moscow: Vserossiiskii tsentralnyi soiuz potrebitelnykh obshchestv, 1922), 40.

[90] Ibid., 46–47.

[91] Ibid., 99. Jasny incorrectly says that Grinevetskii did not have a timetable; see Naum Jasny, *Essays on the Soviet Economy* (New York: Praeger, 1962), 186.

[92] Grinevetskii, *Poslevoennye perspektivy*, 1.

[93] K. Zagorskii, "Obshchaia programma i plan gosudarstvennykh sooruzhenii," *Narodnoe khoziaistvo*, 1919, no. 8: 41.

information. Krug examined in detail the level, type, and distribution of electrified and mechanized industry in the Central Industrial Region and the capacity needed to electrify it.[94] The works of Grinevetskii, Krug, and other members of the heat committee formed the earliest database about electrification and the economy, information that proved invaluable to the GOELRO plan of 1920.

Krug, a founding member of TsES, claimed that the rationale for regional stations was unequivocally rational, economic, and technical, not political.[95] Engineering professor Lev V. Dreier, another colleague of Grinevetskii from student days at the Moscow Higher Technical School, also argued that achieving maximum economic efficiency— again, determined by the engineer—dictated electrification by state-controlled regional stations built with standardized equipment.[96] The promoters of a regional station in White-controlled Rostov-on-Don advocated a similar efficiency-oriented approach to put Russian industry and society on a new footing.[97]

In the creation of political support for electrification, the most important group were the rare hybrid Bolshevik electrical engineers, who worked in Moscow and served as the link between the Communist party and electrification. These men spoke with a technical authority party nonengineers could not match, and they had much better ties with the new leadership than the Petrograd engineers ever had with the tsarist or provisional governments. Lenin's acquaintance with several electrical engineers and technicians stretched back two decades. Krzhizhanovskii and Vasilii V. Starkov shared Siberian exile with him; Krasin knew Krupskaia, Lenin's wife, before Lenin did and could have assumed the leadership of the Social Democrat party in 1907; Klasson and Radchenko met Lenin in 1895.[98] Krzhizhanovskii in particular enjoyed a close relationship with Lenin. These Bolsheviks were only a few of the electrical engineers who promoted wide-scale electrification and staffed the new government offices. Of the approximately 250 people who worked on the GOELRO plan in 1920, only

[94] Krug, *Elektrifikatsiia Tsentralno-promyshlennogo raiona*, 4–37, 40–54.

[95] Ibid., 43–44, 50–51.

[96] Dreier, *Zadachi i razvitie elektrotekhniki*, 11, 20–23; Lev D. Belkind, *Karl Adolfovich Krug* (Moscow: Gosenergizdat, 1956), 11.

[97] *Raionnaia elektricheskaia stantsiia Donetskogo basseina na Lobovskikh Koniakh* (Rostov-on-Don: S. S. Sivozhelezov i Ko., n.d.), 2–3.

[98] Gleb V. Lipenskii, *Moskovskaia Energeticheskaia* (Moscow: Moskovskii rabochii, 1976), 19; Mark O. Kamenetskii, *Robert Eduardovich Klasson* (Moscow: Gosenergoizdat, 1963), 175; Michael Glenny, "Leonid Krasin: The Years before 1917, an Outline," *Soviet Studies* 22 (October 1970): 192–221.

eight were Bolsheviks.[99] What distinguished Bolshevik electrical engineers was their high status in both the Communist party and engineering communities.

According to Thomas Remington's study of socialism and technology, state technologies attracted five types of Communists during this period: inactive Bolsheviks with engineering and management skills (Krzhizhanovskii, Krasin), active Bolsheviks strongly committed to technical progress (Lenin, Aleksei I. Rykov, the VSNKh chairman), Mensheviks who joined the Bolsheviks (Iu. Larin, war communism's "magician of economics"[100]), Mensheviks who did not join (statistician Vladimir G. Groman, who had an "obsession with central planning" and worked for the Main Fuel Committee and Gosplan[101]), and leftist Communists who were fervent technological rationalists (Bolshevik journalist Boris Kushner).[102] Remington also includes scholars interested in central planning though hostile to the new state (Grinevetskii), but he neglects other non-Communist professionals and political activists interested in harnessing technology for economic, political, and social goals. These activists, such as Osadchii and P. I. Palchinskii, an engineer entrepreneur who formed a prewar coal syndicate and served as deputy minister of trade and industry under the provisional government, shared a common interest in a centralized, directed economy, whether from the viewpoint of the engineer's technocratic rationalism, the Menshevik's "technocratic soul," or the Bolshevik's "democratic centralism."[103] At the heart of these interests lay the technology-based goals of the modernization, rationalization, and transformation of Russian society. Bolsheviks constituted a small minority of these technological enthusiasts; ideas of planning and societal transformation by technology were not a Communist monopoly but widely disseminated and shared among engineering and political elites.

Both sides in the civil war had advocates of electrification, but the overwhelming majority remained with the government holding Moscow and Petrograd. If a White government had controlled the two

[99] Cummins, "The Road to NEP," 10, 74–77, 85–86.

[100] Simon Liberman, *Building Lenin's Russia* (Chicago: University of Chicago Press, 1945), 21.

[101] Naum Jasny, *Soviet Economists of the Twenties: Names to Be Remembered* (Cambridge: Cambridge University Press, 1972), 99.

[102] Thomas F. Remington, *Building Socialism in Bolshevik Russia* (Pittsburgh: University of Pittsburgh Press, 1984), 114–16.

[103] Silvana Malle, *The Economic Organization of War Communism, 1918–1921* (Cambridge: Cambridge University Press, 1985), 206.

cities and Red forces waged a civil war, most electrical engineers would have worked under the Whites. Indeed, only the apologia for halting construction of a regional station in White-held Rostov-on-Don distinguished it from stations under Communist control:

> Political events [flow] one after another, like waves, but the Bolshevik wave, accompanied by destruction and plunder, threatens to brake if not stop this project of importance to the state. Only after this numbness passes will the pulse of social-legal peace return for private initiative, will the [company] turn its energy to the project which will play such a large role in the construction of a United Great Russia.[104]

Electrification had a growing base of support in the government. The numerous wartime commissions and other bodies established to regulate fuel and power played an important role in institutionalizing electrical engineers into the leadership circles of the country. This infusion of electrical engineers into government positions did not take firm root until the February revolution but expanded rapidly after the October revolution. The "Moscow mafia" based at the 1886 Company created and filled new state positions, while the heat committee provided important theoretical and factual underpinnings for electrification planning and contributed to immediate utility survival.

Until surpassed in the 1920s by Gosplan, the heat committee dominated regional electrification planning.[105] Its members wrote on electrification, developed the city-saving Makarev boiler, and worked in government bureaus.[106] The committee's most prominent members, Kirsh, Grinevetskii, Krug, and Leonid K. Ramzin, never equaled the 1886 Company's "Moscow mafia" in importance for reasons of politics, death, and institutional weakness. The committee's members tended to the more moderate Menshevik and Cadet political factions, which restricted any possibility of leadership under Bolshevik rule. The deaths of Kirsh and Grinevetskii in the 1919 typhus epidemic eliminated two leaders.[107] Finally, these men worked in an academic environment and did not participate directly in utility and govern-

---

[104] *Raionnaia elektricheskaia stantsiia Donetskogo basseina,* 17.

[105] Glazunov and Sirotinskii, "Uchastie Moskovskogo energeticheskogo instituta," 13.

[106] See Vladimir Kartsev, "The GOELRO Plan and the Emergence of a Science of a New Type," in *Science and Technology: Humanism and Progress* (Moscow: Social Sciences Today Editorial Board, 1981) 2: 41.

[107] Bailes, *Technology and Society,* 53.

ment operations. Ultimately, these professors could advise but not lead.

The transfer of the capital to Moscow in 1918 and the ascendence of the practicing over the university engineer reduced the importance of the Petrograd electrotechnical leadership. Osadchii is an excellent example of the transfer of leadership. Before October 1917, he headed the VI Section, the new Union of Electrotechnicians, and the SED. After October, he worked for the Petrograd branches of the TsES and later GOELRO. In 1921, he moved to Moscow to work for Gosplan, the state planning agency, and the science and technical section of the VSNKh. The shift to Moscow of the electrical engineering leadership symbolized its eagerness to establish closer ties with the new government and the state's recognition of the necessity of electric power for industrial society. Neither electrical engineers nor government could prosper without the other. In contrast, the Academy of Sciences resisted Bolshevik rule and did not move physically to Moscow until 1934 as part of a state centralization of the management of science following a reorganization that reduced the academy's independence.[108]

One striking aspect of Soviet rule was the multiple posts held by engineer-managers, a reflection of the hodgepodge of new organizations, shortages of skilled people, and the extraordinary outburst of technocratic enthusiasm. Krzhizhanovskii, the premier example, in 1919 had executive responsibilities in the KGS, Elektroperedacha, the ETO, the Extraordinary Commission for Electricity Supply in Moscow, the operating board of the Bogorod electric stations, and the board of the state wire and cables factories as well as serving as a delegate to the Moscow Soviet. Krasin and Boris Kushner, a Bolshevik journalist, formed two-thirds of both the ETO governing board and the Extraordinary Commission for Managing the Unified Electrotechnical Industry.[109] Krasin also was a TsES member, served on the VSNKh presidium, chaired the Special Commission to Supply the Red Army, and worked on foreign trade. Occupying multiple posts instead of eliminating overlapping responsibilities allowed stricter centralized control of the growing energy agencies, but it ensured the continuance of a dense bureaucratic thicket.

From the beginning, planning concentrated on the larger concerns

---

[108] Alexander Vucinich, *Empire of Knowledge: The Academy of Science of the USSR (1917–1970)* (Berkeley: University of California Press, 1984), 149.

[109] TsGANKh f. 3429, op. 1, ed. kh., 1162, 135–37, 159; V. Iu. Steklov, *Gleb Maksimilianovich Krzhizhanovskii: Zhizn i deiatelnost* (Moscow: Nauka, 1974), 280; Venediktov, *Natsionalizatsiia promyshlennosti i organizatsiia*, 309.

of wide-scale electrification. At the first session of the TsES in October 1918, chairman Krasin set four immediate tasks: establish nationwide norms for transmission voltages and frequencies, a sine qua non for large-scale networks and for standardized equipment; elucidate the number, power, and distribution of regional stations required to ease the fuel shortage and harness undeveloped natural resources; ascertain the availability of supplies for central stations; and create a general plan for the electrification of Russia.[110] Krasin's first three tasks were specific, but the fourth was an all-encompassing amalgam of other goals—electrifying railroads and industrial regions, establishing and expanding electrotechnical industries, setting standards and examinations, and supplying the needed specialists—into the overall goal of national electrification.[111]

The status of electrification increased significantly from 1917 to 1920. The first congress of regional planning bureaus in May 1918 did not mention electrification. Seven months later, the second congress stated that electrification "can give concrete results only after much time but will instantly place industry on new rails [and] could change the entire nature of our industry."[112] The favorable perception of electrification paralleled the creation of agencies advocating electrification. In May 1918, Elektrostroi, the TsES, and its regional bureaus did not exist; in December, they did.

In a major mid-1919 article, Konstantin Ia. Zagorskii, a railroad economist with an interest in municipalization, called plans for regional stations premature but necessary.[113] Zagorskii wrote to sober planning enthusiasts and temper their utopian declarations. Only rational projects with a possibility of success deserved serious consideration. Although the foremost priority remained transportation, electrification promised, *"like railroads,* to serve as an initial point for wide transformations in very different parts of our social, economic and political life. . . . It is impossible not to conclude that this category of state construction must attract special attention and care from the KGS. . . . It is urgent as quickly as possible to work out a system for the wide and planned physical electrification [*elektrofitsirovanie*] of all regions and areas of the country."[114]

[110] "O pervoocherednykh zadachakh TsES," in Gladkov, ed., *K istorii plana,* 4.
[111] "Raspredelenie voprosov mezhdu chlenami soveta," in ibid., 47–49.
[112] R. Anskii, "Itogi sezda Sovnarkhozov," *Narodnoe khoziaistvo,* 1919, nos. 1–2: 10.
[113] K. Zagorskii, "Obshchaia programma i plan gosudarstvennykh sooruzhenii," 41. For municipalization, see "Khronika gorodskogo dela," *Gorodskoe delo,* 1915, no. 6: 338–39.
[114] Zagorskii, "Obshchaia programma," 44, emphasis added.

Zagorskii outlined the most detailed rationale for electrification yet to appear in the VSNKh journal *Narodnoe khoziaistvo* (National Economy). His broad perspective included the electrification of major industrial regions but also of the countryside, where the vast majority of the population lived. Zagorskii had taken a large step beyond the cities to the full electrification of the country. Regional stations hitherto were viewed mainly as supplying power to cities. Now he redefined the region to benefit the thousands of villages. This line of thought would become more important as the Communist party placed increasing importance on strengthening its rural ties.

Electrifiers did not hold a monopoly on technocratic proposals for economic reconstruction. But electrification differed from railroads and other technologies of the first industrial revolution in its promise of a new tomorrow. In revolutionary times, the visionary initiative went to the technology promising transformation, not simply reconstruction.

Regional Stations

IN CONTRAST with the gloomy state of utility operations, the future of regional stations appeared increasingly bright. By 1920, Lenin's political support had allowed construction to begin on four Soviet regional stations around Moscow and Petrograd. The ideas and siting for these stations were not new; the high-level support was. In 1917–19, Lenin discussed the dismal electrical supply of Petrograd and Moscow with Communist and non-Communist electrical engineers.[115] Together with the work of Elektrostroi, the ETO, and the TsES, these conversations spawned the hydroelectric dams of the Volkhov and Svir rivers for Petrograd and the brown coal Kashira plant and peat-fired Shatura station for Moscow. Although the initial results were minor, the quick Soviet action markedly contrasted with the inaction of the old regime.

Petrograd officials, faced with the most desperate energy situation, had responded quickly to the new government. Before Elektrostroi existed, the Northern Region planning bureau formed an electrification committee, which in March 1918 agreed with the bureau's

---

[115] Besides Krzhizhanovskii, Lenin met with Aleksandr V. Vinter, the future head of Elektrostroi; Genrikh O. Graftio, a long-time hydropower advocate; Piotr G. Smidovich; Ivan I. Radchenko, a peat specialist; and Ivan V. Egiazorov, part of a group studying the Svir. See *SRE*, 24, 29–30, 32, 154; see also Lev B. Kamenov, ed., *Leninskii sbornik* (Moscow: Partiinoe izdatelstvo, 1937), vol. 21, 226–27.

Opening of the Kashira regional station. Courtesy of the Soviet Polytechnic
Museum.

committee on economic policy to construct Volkhov and Svir hydrostations.[116] Elektrostroi quickly assumed the responsibility for both
hydroelectric projects and predicted their completion by 1922–23.[117] At
Lenin's urging, Genrikh O. Graftio, who had formulated tsarist hydropower proposals, headed the electrical engineering section of the
Volkhov project.[118]

[116] "Otchet o deiatelnosti otdela elektrotekhnicheskoi promyshlennosti pri VSNKh,"
Narodnoe khoziaistvo, 1918, no. 5: 30; "Iz polozheniia ob otdele po podgotovke elektri-
fikatsii severnogo raiona," in Gladkov, ed., *K istorii plana*, 62–64; Venediktov,
Organizatsiia gosudarstvennoi promyshlennosti v SSSR, 346.

[117] "Organizatsiia proektirovaniia i stroitelstva elektrostantsii," 29 June 1918, in Gladkov, ed., *K istorii plana*, 41–42; "Iz doklada Vintera," *Krasnyi arkhiv*, 1939, no. 96: 26.

[118] Steklov, *Lenin i elektrifikatsiia*, 16; Genrikh O. Graftio, "Vstrechi," *SRE*, 30. The
model of a bourgeois specialist, Graftio shocked one Communist who met him in 1918

Elektrostroi also controlled the 12-MW Kashira and 5-MW Shatura projects.[119] These stations were salvage operations that relied on existing unused equipment and new imported equipment. The Shatura station was an engineering prototype to determine the feasibility of naval boilers for stationary powerplants. If it was successful, Elektrostroi intended to build a 40-MW station. The turbine for the Shatura station came from the Russo-Baltic factory, and its boilers came from the Provodnik factory.[120] The uncertainty accompanying Kashira was such that the initial designs called for use of Donets coal and Baku oil as well as local brown coal. Blueprints for the turbines and boilers accommodated a wide range of the possible available equipment.[121] The technical challenge was the development of a boiler to burn brown coal efficiently. Because of its closer proximity to Moscow, a nearby river and railroad, and prior construction, Kashira received a higher priority than Shatura, with initial operations optimistically predicted by 1920.

Russian electrical engineers broke new ground with these stations. The construction of the Volkhov and Svir stations, the first large-scale Russian hydrostations, demanded mastery of new concepts and technologies. All four stations used long-distance, high-voltage (115 kV) transmission lines, another area of limited experience. These technical challenges, combined with financial, material, and personnel shortages, ensured that Elektrostroi's optimistic schedules would not be met.

These stations' most novel aspect was not the technical challenges—Western countries had met them—but the political support for regional stations and the underlying concepts of wide-scale electrification and economic transformation. A 1920 *Narodnoe khoziaistvo* article on industrialization called "measures of a technical character to change the very structure and conditions of production" a priority behind only transportation.[122] As in 1918, localizing production and consumption took precedence, but now new materials, fuels, and technical processes were perceived as ways to develop local substi-

---

by his formal manner, formal clothing, and reference to "Mr." (*grazhdanin*) Lenin; see N. P. Bogdanov, "Skvoz grozy i buri," ibid., 43.

[119] "Organizatsiia proektirovaniia i stroitelstva elektrostantsii," 29 June 1918, in Gladkov, ed., *K istorii plana*, 41–42.

[120] "Khronika," *Ekonomicheskaia zhizn*, 12 April 1919, 3; "Iz doklada Vintera," *Krasnyi arkhiv*, 1939, no. 96: 28.

[121] M. L. Kamentskii, "Pobedenskaia raionnaia elektricheskaia stantsiia," *Voprosy elektrifikatsii*, 1922, nos. 1–2: 100–101.

[122] N. Charnovskii, "Znachenie raionirovaniia promyshlennosti," 17.

tutes. The needs were the same but a new rubric was used, that of new technologies to achieve wider economic and social goals. And electrification was the new technology par excellence.

The October revolution occurred at a paradoxical point in the development of Russian electrification. At the start of 1920, the present never appeared so appalling, nor had the future ever looked so bright. The problem was the transition from the dismal present to a radiant future. The utilities were in far worse condition than in 1917: the Petrograd utilities, once the pride of Russia, burned wood and produced only a third of their 1916 output. Stations elsewhere suffered from equally serious shortages and the decay of the infrastructure.

The war had demonstrated the link between electric power and industrialization. Now the importance of electricity expanded to encompass the transformation of industry and society. A network of political, economic, and engineering actors began coalescing as the vision of electrification diffused and electrical engineers established footholds in executive and planning sections of the new government. The attraction of an electrically transformed future spread from the electrical engineers to other economic and political elites and attracted mass publicity as part of a larger wave of technological and revolutionary utopianism. New forward-looking organizations of electrical engineers, of utilities, and, most important, in the government established an organizational framework for electrification even as the boundaries between utilities and the state collapsed.

The interest in electrification grew not only from its intrinsic attractions but also from the devastation of the existing technical infrastructure, especially the railroads. Regional electrification by locally fueled stations fit the autarkic economic and political conditions of revolutionary Russia. The growing political and engineering desire to use technology to transform society also increased the attraction of electrification, which had gained a political focus: it became the connecting point between planning for rational economic development and large-scale remolding of the social and industrial landscape.

CHAPTER 6

GOELRO: The Creation
of a Dream, 1920–1921

In December 1920, the Communist party made electrification the new state technology by approving the GOELRO plan. The creation of this state plan of electrification took ten months; final approval in a modified form demanded another year. Instead of a future desideratum, electrification became the most important and immediate way, together with planning, to reconstruct the economy and modernize the country.

Credit belongs to the entrepreneurial drive of several electrical engineers, especially Gleb Krzhizhanovskii, who forged an actor-network that created allies and promised resources for state electrification. They persuaded an already interested Lenin to form a government commission to propose a state network of regional power stations. Once established as GOELRO (State Commission for the Electrification of Russia), these engineers expanded their mandate to create the first comprehensive industrial plan for Soviet Russia. A vacuum about the country's future course existed, and this group of electrification advocates, fashioning a political alliance with the Communist party, filled it.

The GOELRO plan benefited both the Communist party and the electrical engineering community. Electrification formed the foundation of Bolshevik plans for economic and societal reconstruction: it would eliminate the town–country split, create a communist society in Russia, and lead to a similar transformation of the capitalist world while the electrical engineers received official support for their utopian visions.

A good analogy is Ronald Reagan's decision to create the "Star

Wars" Strategic Defense Initiative in March 1983.[1] In both cases, a technical adviser and confidant—Edward Teller, Gleb Krzhizhanovskii—pushed utopian plans the leader already had an interest in; the necessary ideas and institutions already existed; the technologies had not actually been deployed; both leaders invested much political capital; and neither plan survived criticism intact. This analogy may shock—Lenin and Reagan are rarely linked—but it is not unwarranted. States do not adopt technologies in vacuo; such actions occur within a pattern including prior state interest in the technology, a politically connected engineering or scientific entrepreneur, a ruling party facing a perceived challenge or crisis, and a political leader who promotes the technology for specific political goals.

Electrification in Soviet Russia fits this pattern. In 1919–20, civil war and the failure of the railroad system severely damaged the economic and political order. The GOELRO plan promised a future society remade in a desired way for a desired end. It fit Howard P. Segal's definition of a technological utopia: "The use of hardware (structures and machines alike) and, in addition, of knowledge (technical and scientific alike) to create and preserve an avowedly perfect society."[2] GOELRO reflected the enthusiasm and hopes not only of Bolshevik and non-Bolshevik electrical engineers but of those who saw a revitalized Russia rising from the devastation of war and revolution. The electrical engineers promoted their most optimistic visions, but these were visions shared and supported by many others. What distinguished the electrical engineers was their ability to bring heterogeneous political, technical, and economic elements together.

Yet the focus on regional stations was not inevitable, nor was the ascent to state technology without complications. Electrical engineers had a choice of three technological paths for post–civil war electrification: a centralized path of building regional stations, a conservative path of expanding existing utilities, and a radical path of rapidly electrifying the countryside. In 1920, GOELRO chose the first path without seriously considering the others. To some, this choice demonstrated the triumph of hope over reality. To others, GOELRO was an opportunity lost, a continuation of tsarist industrialization instead of socialist transformation.

---

[1] David Perlman, "The Dilemmas of Decision-Making: From AIDS to Star Wars," in William T. Golden, ed., *Science and Technology Advice to the President, Congress, and the Judiciary* (Elmsford, N.Y.: Pergamon, 1988), 257.

[2] Howard P. Segal, *Technological Utopianism in American Culture* (Chicago: University of Chicago Press, 1985), 14.

By October 1921, advocates of the alternatives had started to form their own actor-networks with critics of the centralized GOELRO plan, which set the stage for the political struggles over the future of electrification through the mid-1920s. The growth of technical challenges to the plan corresponded to major political changes between the establishment of the commission in February 1920 and the plan's full approval in December 1921. The shift from war communism to the NEP upset the political and economic foundations of the GOELRO plan, which included the resumption of economic and diplomatic relations with the West and a supportive domestic front. The challenges to GOELRO may have been phrased technically, but they concerned the very nature and direction of the Soviet state. Would it be directed from the center or guided from below? The choice of electrification embodied in GOELRO strengthened the first direction.

The adoption of the GOELRO plan contrasts with the lack of acceptance of similar schemes in the West. Why did Soviet Russia accept the utopia of electrification before the West, previously Russia's example? The answer lies in the very different environments in Russia and the West, but also in the coalition forged by Krzhizhanovskii and Lenin.

Lenin's Role in Electrification

VLADIMIR ILICH LENIN's eager advocacy raised electrification to its new prominence. Without Lenin, the government probably would have adumbrated an electrification plan, but neither as large nor with the same evangelical enthusiasm and political support. Part of a lifelong fascination with technology to solve social problems, Lenin's interest in electrification dates to his 1896 *Razvitie kapitalizma v Rossii* (Development of Capitalism in Russia), written during Siberian exile with Krzhizhanovskii.[3] By 1914, Lenin envisioned a future in which "the 'electrification' of all factories and railways [would] accelerate the transformation of dirty, repulsive workshops into clean, bright laboratories worthy of human beings," and household electric light and heating would ease the life of millions of "domestic slaves." Most important, electricity would eradicate the cultural and economic chasm between town and country, a major target of Russia's Marxist modernizers dating from Engels's 1883 optimistic letter about the

---

[3] Vasilii I. Steklov, *V. I. Lenin i elektrifikatsiia* (Moscow: Nauka, 1975), 19; Adam B. Ulam, *The Bolsheviks* (New York: Collier, 1973), 134, 457.

value of electricity.[4] "We must show the peasants that the organization of industry on the basis of modern, advanced technology, on electrification which will provide a link between town and country, will make it possible to raise the level of culture in the countryside and to overcome, even in the most remote corners of the land, backwardness, ignorance, poverty, disease, and barbarism."[5]

Like many nonengineers and Marxists, Lenin held overly optimistic and utopian opinions of the possibilities of science and technology. As Adam Ulam puts it,

> In his dream of science as the magic key to the future Lenin was both a faithful disciple of Marx and at one with that sizable portion of the Russian intelligentsia who, with Gorky, had despaired finally of purely political solutions and saw in technology the most hopeful way of civilizing "the half-savage, stupid and heavy" common man. . . . The words "after the electrification of Russia" now assume in Lenin's mouth the same role that the words "after the Revolution" played prior to October. Everything will be different.[6]

After meeting Lenin in September 1920, H. G. Wells expressed a similar opinion: "For Lenin, who like a good orthodox Marxist denounces all 'Utopians', has succumbed at last to a Utopia, the Utopia of the electricians."[7]

Lenin proved invaluable in prodding the state bureaucracies into action. His collected works contain numerous letters to obtain supplies, gather information, and make disagreeing officials agree.[8] His assistance proved especially helpful in obtaining foreign equipment for regional stations, an assist necessitated by the newly established state monopoly on foreign trade.[9] Intervention from above should be

[4] V. I. Lenin, "A Great Technical Achievement," *Collected Works*, 4th ed., ed. Institute of Marxism-Leninism, 60 vols., (Moscow: Progress Publishers, 1964), 19: 61–62; "Four Thousand Rubles a Year and a Six-Hour Day," ibid., 20: 68–70; "Agrarian Question and the 'Critics of Marx,'" ibid., 5: 140. See also Richard Stites, *Revolutionary Dreams: Utopian Vision and Experimental Life in the Russian Revolution* (New York: Oxford University Press, 1989), 47, 52; Steklov, *Lenin i elektrifikatsiia*, 15.

[5] Lenin, "Report on Work of All-Russia C.E.C. and C.P.C.," 2 February 1920, *Collected Works*, 30: 333.

[6] Ulam, *Bolsheviks*, 481.

[7] H. G. Wells, *Russia in the Shadows* (New York: George H. Doran, 1921), 158–60.

[8] E. g., see Lenin, *Collected Works*, 35: 462–63; 45: 75–76, 115–17, 273–74, 464, 494, and 501. See also Lev B. Kamenov, ed., *Leninskii sbornik* (Moscow: Partiinoe izdatelstvo, 1952), 20: 209–21.

[9] See, e. g., Viktor S. Kulebakin, "Skromnyi vklad," in *SRE*, 107; John Quigley, *The Soviet Foreign Trade Monopoly* (Columbia: Ohio State University Press, 1974), 58–59.

seen not just as Lenin's personal interest but as indicative of the chaotic and turgid nature of Soviet bureaucracy, a direct successor of the tsarist bureaucracy. Projects without Lenin's attention fared less well, especially if they fell outside his immediate interests. Other large-scale projects with tsarist roots, such as the Volga–Don canal, had to wait for Stalin, and the state bureaucracy did not allow any project to escape its grasp.[10] As a state technology, electrification benefited from elite backers to obtain resources and ease the constrictions of the government bureaucracy.

Contemporary engineers and subsequent Soviet historians praised Lenin as an "unquenchable propagandist" whose "biggest support" lay in convincing the party and country to back GOELRO.[11] Lenin understood well and used the political and propagandistic value of electrification. The classic example is his December 1920 letter to Krzhizhanovskii about the possibilities of melting church bells for copper and placing a light bulb in every village.[12] The two suggestions were politically astute, but the latter was a task well beyond contemporary resources. Over the decades, the "lamp of Ilich" has served as effective propaganda for the workers' state.[13]

The electrical engineers needed Lenin—but he needed them. Utopian dreams need utopian technologies, and utopian dreamers need political support. After Lenin's death in 1924, an obituary in *Elektrichestvo* stated, "Only 'Ilich' understood the might and role of electricity in the national economy and transformed it from a narrow, technical idea to the ideal of peasants and workers, connecting it organizationally to Soviet power."[14] Lenin was not alone in his entrancement by the promise of electricity nor in his desire for a transformed future, but in the Soviet Russia of 1920 he was the only one who could put the state behind the promise. Krzhizhanovskii's success with the GOELRO plan lay partly in his ability to harness the resources of Lenin—and, through him, the Communist party—to the cause of electrification.

[10] Thomas F. Remington, *Building Socialism in Bolshevik Russia: Ideology and Industrial Organization, 1917–1921* (Pittsburgh: University of Pittsburgh Press, 1984), 124–26, 130, 150–60.

[11] N. P. Bogdanov, "Skvod grozy i bury," *SRE*, 45; Steklov, *Lenin i elektrifikatsiia*, 57.

[12] Lenin, *Collected Works*, 35: 467–68.

[13] The image of the peasant seeing his first light bulb has been immortalized on Soviet lacquer boxes, posters, stamps, photographs—on anything that would convey the message.

[14] A. Z. Goltsman, "Vladimir Ilich Lenin," *Elektrichestvo*, 1924, no. 1: ii.

Soviets and electrification are the basis of a new world. Courtesy of the Hoover Institution.

The Creation of GOELRO

GOELRO ORIGINATED at a 26 December 1919 meeting between Lenin and Krzhizhanovskii to discuss Moscow fuel shortages.[15] From that meeting, one of several between the two, came the organizational seed that sprouted into a national plan for the country's high-technology reconstruction.[16]

At the seventh convocation of the All-Russian Central Executive Committee on 2–7 February 1920, Lenin called for the VSNKh and the Commissariat for Agriculture to draft an electrification plan to provide "a new technical foundation for new economic development."[17] Approving Lenin's request, the executive committee declared, "For the first time Soviet Russia is offered the possibility to start on a more planned economic construction, on the scientific creation and subsequent implementation of a state plan for the whole economy."[18] The resolution called for a list of projected electric stations within two months as one part of a state plan for the entire economy. The huge effort actually lasted ten months and produced not a section of a state plan but a decade-long national plan based on electrification. The initial, muted reporting hailed regional stations and long-distance transmission as saviors of industry; planning was not mentioned.[19]

The electrical engineers quickly organized. The ETO convened representatives of twelve electrotechnical groups on 11 February in Moscow to discuss the electrification of Russia and to urge the "indispensable" creation of a centralized main electrotechnical committee.[20] At their second meeting a week later, these engineers selected eight people with wide institutional support to lead the commission (see Table 6.1). Although Petrograders such as Osadchii and Shatelen participated, GOELRO marked the new dominance of the Moscow leadership under Krzhizhanovskii. The ETO head said that Lenin, who was "carefully following what we do," considered the commission "one of

---

[15] G. M. Krzhizhanovskii, "Perspektivy elektrifikatsii," *Planovoe khoziaistvo*, 1925, no. 2: 4; also Steklov, *Lenin i elektrifikatsiia*, 42.

[16] For a more detailed history, see Alex G. Cummins, "The Road to NEP, the State Commission for the Electrification of Russia (GOELRO): A Study in Technology, Mobilization and Economic Planning" (Ph.D. diss., University of Maryland, 1988).

[17] Lenin, *Collected Works*, 30: 334–35.

[18] "Rezoliutsiia ob elektrifikatsii Rossii," *Biulleten GOELRO*, 1920, no. 1: 2.

[19] P. K-i, "Elektrifikatsiia sovetskoi Rossii," *Bednota*, 7 February 1920, 3; "Elektrifikatsiia Rossii," *Izvestiia Elektrotresta* 1 (March 1920): 3, and 2 (May 1920): 5.

[20] "Protokol soveshchaniia po voprosu ob elektrifikatsii Rossii," 11 February 1920, in Viktor S. Kulebakin, ed., *Trudy GOELRO* (Moscow: Izdatelstvo sotsialno-ekonomicheskoi literatury, 1960), 177.

Table 6.1. GOELRO leaders, 17 February 1920

	Organization
G. M. Krzhizhanovskii[a]	ETO
A. G. Kogan[a]	TsES
B. I. Ugrimov[a]	Commissariat for Agriculture
G. O. Dubelir	Elektrostroi
G. O. Graftio	Section for railroad electrification
K. A. Krug	Heat committee
M. Ia. Lapirov-Skoblo	VSNKh science-technical section
B. E. Stunkel	Glavtekstil (state textiles)

Source: "Protokol zasedaniia," 17 February 1920, *Trudy GOELRO* (Moscow, 1960), 92.
[a] Presidium member.

the most important state organs [,which would] receive the most nourishing support from the state. . . . It seems to me, comrades, that such a relation with comrade Lenin guarantees us a very favorable situation for our work."[21]

Despite the promised support of Lenin, the VSNKh presidium created GOELRO on 21 February 1920 and not a main electrotechnical committee.[22] Under Lenin's prodding, GOELRO received a budget of 20 million rubles and the right to communicate directly with other government organs because of its state importance and extraordinarily urgent work. The 9th Party Congress, which met in late March and early April, strengthened GOELRO's position by calling for the creation of an electrification plan as part of a single economic plan.[23]

GOELRO did not deliver its final report until December, eight months behind schedule. A lack of oversight contributed to the delay as the Polish war diverted the country's leadership, and GOELRO planners used the time to expand their mandate. Project delays are not uncommon and are often linked to efforts to redefine, restructure, and clarify problems, obtain more information, and negotiate with interested communities.[24] GOELRO suffered from all these normal

[21] "Iz stenogrammy Soveshchaniia komissii po elektrifikatsii," 17 February 1920, in "K istorii elektrifikatsii RSFSR," *Krasnyi arkhiv*, 1939, no. 95: 32–33.

[22] "Vypiska iz protokola zasedaniia," *Biulleten GOELRO*, no. 1: 2.

[23] "Polozhenie," ibid., 3; *KPSS v rezoliutsiiakh i resheniiakh sezdov, konferentsii i plenumov Ts.K.*, 7th ed. (Moscow: Politizdat, 1953), 478–79; Cummins, "Road to NEP," 176–79; D. Baevskii, "Leninskii plan sotsialisticheskogo preobrazovaniia Rossii i GOELRO," *Voprosy istorii*, 1947, no. 3: 11–13.

[24] See, e. g., Joan H. Fujimura, "Constructing 'Do-able' Problems in Cancer Research: Articulating Alignment," *Social Studies of Science*, 17 (1987): 257–93.

problems but also from several factors unique to Russia in 1920. Foremost was the limited number of trained people. For many, GOELRO served as a second or third job, limiting their involvement.[25]

Nor could "these ragged engineers, working in rooms which they can hardly keep above freezing-point and walking home through the snow in boots without soles" take food, housing, and health for granted.[26] At least one GOELRO member, professor Viktor S. Kulebakin, spent early 1920 recovering from typhus, the killer of Grinevetskii and Kirsh in 1919.[27] To avoid the famine that killed, among others, seven of the forty-five members of the Academy of Sciences, GOELRO workers and their families attempted to receive Red Army rations instead of the lesser civilian rations.[28] The Petrograd section suffered acutely from fuel shortages.[29] The VSNKh had to intervene so that Aleksandr Kogan, a member of the GOELRO presidium, could have a third room in his apartment.[30] The appalling conditions in which GOELRO operated make its work that much more impressive.

Officially, the ETO controlled GOELRO, but it acted independently. GOELRO depended heavily on the new electrification infrastructure and especially the TsES, whose subsections became the eight GOELRO regional groups.[31] GOELRO received requests from cities for assistance to build power plants, to expand existing stations, and to cooperate on regional electrification plans.[32] Save for the last, GOELRO passed these requests to the ETO. GOELRO tried to control such independent initiatives with fair success.[33]

Defining GOELRO's task proved more difficult than anticipated.[34] Only in mid-March, a few weeks before the original deadline, did the commission adopt a work outline. GOELRO concentrated on Russia's industrialized, urbanized regions and not on less developed rural

---

[25] Kulebakin, "Skromnyi vklad," *SRE*, 104.

[26] Arthur Ransome, *The Crisis in Russia* (New York: B. W. Huebsch, 1921), 170–71.

[27] Kendall Bailes, *Technology and Society under Lenin and Stalin* (Princeton: Princeton University Press, 1978), 53; Kulebakin, "Skromnyi vklad," *SRE*, 103.

[28] Vladimir N. Ipatieff, *The Life of a Chemist* (Stanford: Stanford University Press, 1946), 271; *Trudy GOELRO*, 107.

[29] 15 May protocol, TsGANKh f. 5208, op. 1 (2), ed. kh. 1, 59.

[30] Thereby allowing him "to lead a normal life"; TsGANKh f. 3429, op. 1, ed. kh. 1953, 8.

[31] "Protokol," 24 February 1920, *Trudy GOELRO*, 100–103; "Glavelektro, Petrogradskoe otdelenie," *Izvestiia Elektrotresta*, June–July 1920, no. 3: 11–12.

[32] E. g., from Rzhev, 16 March, *Trudy GOELRO*, 118; Astrakhan, 28 March, ibid., 122; Rostov, 23 November, ibid., 191.

[33] 30 September, ibid., 174–75.

[34] M. A. Smirnov, "Kak rabotala Komissiia GOELRO," *Elektrichestvo*, 1940, no. 12: 8.

areas because only large industrial users and large cities could provide the high loads necessary to justify the colossal national expenditures.[35] Electrification by regional stations would transform the economy—but only in the developed parts. The countryside would have
to wait. Inherent in the focus on regional stations, this flaw in the
GOELRO plan would lead to alternative proposals in 1921 to electrify
the countryside rapidly by decentralized, small-scale local stations.

Following the Social Democratic tendency to distinguish between
the immediately practical and revolutionarily desirable, GOELRO divided its activities into a short-term "minimum" and a long-term
"maximum" program.[36] The former sought to increase quickly the use
of existing utilities and factory powerplants.[37] The construction of
transmission networks for these stations would pave the way "for the
regulation and correct socialization of the entire electrical economy."[38]
The maximum program, the heart of the GOELRO plan, focused on
the creation of a state network of regional stations. It divided Russia
into eight geographic regions (*raiony*), each with its own GOELRO
group to survey existing and potential economic development, transportation, resources, and electrification's role over the next decade.[39]
Although the roots of this geographic approach lay in tsarist and
pre-1920 efforts, GOELRO was the first planning organ to approach
regional economic development from a national and not local basis.[40]
The quality of work varied by region because of differences in data,
their processing, and their analysis. The analysis of the Central Industrial Region was the most technically competent and comprehensive.
Some regions produced abundant statistics; others, such as Tur-
kestan, provided only glimmers of coverage, reflecting the poor development of both industry and government.

From these surveys, GOELRO's plan for economic development,
based on a standard methodology, emerged. According to Krzhizhanovskii, the priorities for development were to reconstruct the nation's economic base, manufacture machinery to increase industrial

[35] K. A. Krug, *Programma rabot po elektrifikatsii Rossii* (Moscow: Trudy GOELRO, 1920),
4.

[36] E. H. Carr, *The Bolshevik Revolution, 1917–1923*, vol. 2 (New York: Macmillan, 1951),
9.

[37] "Protokol," 28 February 1920, *Trudy GOELRO*, 104.

[38] "Programma rabot i poiasnitelnaia zapiska k nei," *Biulleten GOELRO*, 1920, 1: 6.

[39] Ibid., 8–10.

[40] Gleb M. Krzhizhanovskii, *Voprosy ekonomicheskogo raionirovaniia SSSR: Sbornik materialov i statei (1917–1929 gg.)* (Moscow: Gosudarstvennoe izdatelstvo politicheskoi literatury, 1957), 3–5.

output, and produce goods for mass demand.[41] Underlying these priorities was the assumption that an active policy and state creativity in industrialization would change Russia from a colonial market of Europe to an independent power.[42] The prescription is now familiar: industrialize first, satisfy consumer demand later. These politically statist priorities embodied a centralized technocratic bias that stood chronologically and ideologically in the continuum between the trickle-down industrial policies of tsarist minister Sergei Witte and Stalin's superindustrialization.

Three themes intertwined throughout the GOELRO analyses. Foremost was a reliance on foreign technology and capital. Second was slicing through the Gordian knot of the transportation-fuel-food crises. Last was an avoidance of conventional thermal stations. If hydropower was not possible, thermal plants would use low-quality local fuels, ranging from peat to industrial wastes.

One important component of GOELRO's technocratic mindset was the determination of the appropriate units of analysis, as its approach to railroads and planning indices demonstrates. One difference between the old state technology, railroads, and the new state technology, electrification, was the latter's more encompassing conceptual framework. When the railroad industry, which traditionally took a long-term view, presented its ten-year plan, GOELRO found the plan lacking a firm economic basis, divorced from other forms of transportation, and based on poor assumptions about fuel. Although Krzhizhanovskii approved the industry's underlying thesis of building railroads for foreign trade, he stated that the unit of analysis should be networks, not individual lines.[43]

GOELRO wrestled with the question of appropriate planning indices for Russia's shattered economy. Due to hyperinflation caused by the collapse of the ruble and the return to a barter economy, monetary data were next to useless, and other economic indicators were hard to obtain.[44] This destruction of the old economic order supported efforts by technocrats and socialists to replace "the antiquated notion of the market cost of labor."[45] In a significant analytical leap in scale

[41] 28 March, *Trudy GOELRO*, 122.

[42] TsGANKh f. 3429, op. 1, ed. kh. 1953, 64.

[43] "Protokol," 5 June, TsGANKh f. 5208, op. 1 (2), ed. kh. 1, 77b-80b; also, *Trudy GOELRO*, 153.

[44] Using 1913 as the base, prices increased by a factor of 55 in 1918, 500 in 1919, and 100,000 in 1920; see I. A. Skavani, "Elektrosnabzhenie Leningrada," *Elektrichestvo*, 1924, no. 4: 182.

[45] TsGANKh f. 3429, op. 1, ed. kh. 1953, 52b, 72.

and approach from earlier efforts, GOELRO employed standardized concepts and methodologies, including material and fuel balances, which greatly shaped future Soviet statistics.[46] This work proceeded independently of another VSNKh commission working toward similar goals.[47]

GOELRO expressed great interest in material indices, already used in the United States for placement of industries based on location of resources, markets, and transportation. Of special importance was Krzhizhanovskii's concept that planning incorporate an energy analysis of economic processes "as the most opportune key to settle the vexed questions of economic construction," a concept that also fascinated Henry Ford.[48] Boris Kushner, a director of the nationalized electrotechnical industry, extended the logic of electrification as the technology of the new age in 1920 when he proposed the kilowatt-hour as an index of culture and progress.[49] Krzhizhanovskii suggested an economic accounting unit based on the amount of manpower needed to produce a given quantity of wheat, a concept more appropriate to agrarian Russia.[50]

Actual application of these ideas came in the pioneering work of Leonid K. Ramzin and Krug. Ramzin, a thermal engineer and professor at the Moscow Higher Technical College, calculated the energy supply and demand of cities and regions, but he was too thorough for GOELRO. Although he provided highly valued material, according to Krzhizhanovskii, Ramzin's interpretations produced heated debate and opposition within the commission.[51] His coverage encompassed the total fuel balance of Russia and thus buried industrial and urban consumption under peasant usage. According to Ramzin's calculations, vegetable fuels—wood, straw, and manure—had provided between 60 and 80 percent of Russia's entire fuel consumption since

[46] G. Koginov, "Statistika v plane GOELRO," *Vestnik statistika*, 1950, no. 6: 3–14; Eugene Zaleski, *Planning for Economic Growth in the Soviet Union, 1918–1932* (Chapel Hill: University of North Carolina Press, 1971), 38.

[47] Silvana Malle, *The Economic Organization of War Communism, 1918–1921* (Cambridge: Cambridge University Press, 1985), 190–93.

[48] G. M. Krzhizhanovskii, "Tekushchie voprosy elektrifikatsii," *Elektrichestvo*, 1922, no. 2: 4. Krzhizhanovskii here criticized a 1922 article in *Zeitschrift des Vereines D.I.* by Klingenberg and Ridler on the future energy economy of Germany. See also David A. Hounshell, *From the American System to Mass Production, 1800–1932* (Baltimore: Johns Hopkins University Press, 1984), 281.

[49] Boris Kushner, *Revoliutsiia i elektrifikatsiia* (Petersburg: Gosudarstvennoe izdatelstvo, 1920), 7.

[50] TsGANKh f. 3429, op. 1, ed. kh. 1953, 52b, 72.

[51] Ibid., 75 and 76b, 104.

1909.[52] From this viewpoint, electrification's contribution to the economy could be only minor—definitely not the conclusion desired by GOELRO. Ramzin warned against electrification as a panacea, for "electrification is a means for a very small increase in the heat balance of the country, and it alone will not solve the fuel question."[53]

In cautioning against a misplaced faith in the inherent superiority of local fuels, Ramzin touched another soft spot in GOELRO's plan: when "better" fuels became available as the economy improved, would use of local fuels continue? The student of Grinevetskii warned that market choice favored the fuel with the highest energy content and lowest cost, that is, oil or coal. With the exception of hydropower, utilities favored local fuels only because better fuels were not easily available or affordable.[54]

Ramzin employed a more encompassing conceptual framework than that used by GOELRO. Echoing Krzhizhanovskii's contention that changing the habits of 150 million peasants was utopian, GOELRO maintained its urban focus on industrial and not total fuel use. Ramzin's view of electrification as a subordinate factor was, according to Krzhizhanovskii, "a misunderstanding and differs from the opinion of professor V. I. Grinevetskii."[55]

Rural Electrification

INDICES AND ENERGY balances typlified the comprehensive thinking of the technocratic electrical engineer. As the problems GOELRO encountered with Ramzin's larger framework demonstrate, however, these urban engineers were unable to comprehend the very different world of Russian agriculture. Feeding the country, providing grain for export, and turning peasants into full members of the socialist state were important, if conflicting, state agricultural priorities.[56] GOELRO considered electrification an essential component of agricultural modernization,[57] but the blinders of high technology for transformation hindered its appreciation of more incremental development, such as

[52] P. S. Neporozhnii, ed., 50 let Leninskogo plana GOELRO: Sbornik materialov (Moscow: Energiia, 1970), 142. For a detailed analysis, see L. K. Ramzin, "The Power Resources of Russia," in W. R. Douglas Shaw, comp. and ed., Transactions of the First World Power Conference (London: Percy Lund Humphries, 1924), vol. 1: 1251.

[53] TsGANKh f. 3429, op. 1, ed. kh. 1953, 76.

[54] Ibid., 76–76b.

[55] Ibid., 74b-75, 105.

[56] Malle, Economic Organization of War Communism, 396–99.

[57] "Polozhenie o GOELRO," Biulleten GOELRO, 1920, 1: 2.

improved hand tools. Similarly, the commission's enthusiasm for regional stations led to a neglect of smaller stations that also would have produced more immediate results. Unfortunately for the rural sector, its urban-based brethren defined the road to utopia too narrowly.

GOELRO encountered opposition from four elements: agricultural specialists and bureaucrats, electrification enthusiasts, electrical engineers, and peasants. Resistance from the Commissariat for Agriculture led GOELRO to seek its data and allies elsewhere. The enthusiasts, who combined great zeal with little rural experience, ignored practical, small steps to concentrate on the total transformation of farming. Many electrical engineers saw agricultural electrification primarily as a means of increasing load factor; for them, the technical challenges existed in the regional stations. Last and least were the peasants themselves, notoriously resistant to change and completely isolated from GOELRO's activities. Because GOELRO failed to establish a commonality of goals and activities among these key groups, a cohesive rural electrification program did not develop.

Before GOELRO, agricultural electrification fell under the purview of the bureau for electrification of the Commissariat for Agriculture, whose initial goal was to substitute electrical energy for animal power.[58] Director Boris I. Ugrimov, who had a long-standing interest in electrification, quickly realized that the bureau's major task should be one of promoting the idea of electrification among uneducated peasants, educated agronomists, and agricultural engineers.[59] This new goal reflected an understanding that the electrifiers had to convince actors at all levels of the agricultural hierarchy of the need and importance of electrification.[60]

Ugrimov headed GOELRO's group on agriculture, which in May outlined its vision of the future.[61] Geographically, electric lights and appliances would diffuse first to farm plots and gardens outside cities and use existing stations. Eventually electrification would transform not just farming but food processing, livestock, and other agrarian activities. Applications included pumps for irrigation, lighting, mo-

[58] B. N. Knipovich, *Ocherk deiatelnosti NKZem za tri goda (1917–20)* (Moscow, 1920), 27, cited in *Trudy GOELRO*, 27.

[59] Boris Ugrimov and his brother Aleksandr, the president of the Moscow Society for Agriculture, and Krzhizhanovskii discussed agricultural electrification with Lenin in May 1919; see A. I. Ugrimov, "Moi put i rabota v GOELRO," *SRE*, 83–92.

[60] "Protokol zasedaniia," *Trudy GOELRO*, 94–96.

[61] Ibid., 114, 128–33.

tors, plows, and even soil electrolysis. The most interesting proposal was to establish 120 experimental test stations to discover the best local uses of the new, mechanized agriculture, act as regional popularization centers for electrification, and provide electricity for the surrounding environs. Distributed nationwide, these stations would give each region a stake in electrification and provide propaganda for electrification.[62]

GOELRO and the Commissariat for Agriculture inhabited very different worlds of knowledge and interests. The unwillingness of the ministry to assist GOELRO helps explain why the electrification of agriculture did not advance significantly. Ugrimov, it should be remembered, targeted electrification propaganda at both technical specialists and peasants. Unlike the industrial sector, the rural sector, including officials at local and state levels, lacked a widespread knowledge of electricity. The state of Russian agriculture contributed to the commissariat's apathy. The technical level of farming and food production was low, distances were vast, farms were too small to mechanize profitably, and the peasants thought of wealth in terms of cattle, not capital. Electrification was almost too advanced for agriculture. An April GOELRO report noted that other forms of mechanization and an increase in the number of horses offered greater return on investment than electrification. The full application of electrification to agriculture would prove an unrealizable dream unless massive transformations independent of electric power swept the land.[63] Electrification seemed sufficient but not necessary for the socialist transformation of rural Russia.

From an engineering viewpoint, what young electrical engineer would want to work on a small plant in rural isolation instead of at a large station with greater professional challenges? Most electrical engineers viewed agriculture as a secondary priority, since the maximum economic benefits of electrification lay elsewhere.[64] Those engineers who wanted to electrify the countryside faced the self-imposed and essentially unavoidable problem of overenthusiasm. A bit of fanaticism aids in overcoming opposition and entering new areas of application, but it does not necessarily produce desired results, as the electric plow demonstrated.[65] A technically viable alternative to steam

[62] TsGANKh f. 5208, op. 1 (2), ed. kh. 1, 6–7.

[63] TsGANKh f. 3429, op. 1, ed. kh. 1953, 17–23; TsGANKh f. 5208, op. 1 (2), ed. kh. 1, 3.

[64] TsGANKh f. 5208, op. 1 (2), ed. kh. 1, 66, 75.

[65] Another idea that never germinated was soil electrolysis, one of the "completely

plows in prewar Germany, electric plows failed because of larger problems of mechanization and social organization of German agriculture.[66] In Russia, knowledge of German efforts and Lenin's interest and support resulted in the continued investment of expertise and expense in developing an electric plow.[67] Most electrical engineers were skeptical or openly hostile, viewing it as a useless expense of time and labor and a fantasy.[68] A few, however, saw such a machine as the perfect exemplar of bringing electricity to the masses, perfectly feasible in ten to fifteen years, and along with electric irrigation pumps a great improvement for production.[69] None fully understood the significant social and organizational changes in agriculture the plow's adaptation would have demanded. Under the more market-oriented conditions of the mid-1920s, research on the electric plow quietly ceased.[70]

Agriculture provoked one of the few ideological disputes within GOELRO. L. N. Litoshenko, an ex-Cadet agricultural economist,[71] concluded that large individual farms offered the best opportunities for electrification.[72] Concerned that he did not endorse state farms (*sovkhozy*), Krzhizhanovskii attacked Litoshenko "from the viewpoint of technology, economics, and understanding the destiny of agriculture" and of deviating from Ballod's assumptions, which greatly influenced GOELRO's conceptions of agriculture.[73] This was the only major ideological dispute in the GOELRO protocols, which contain criticism of specific projects but rarely of political assumptions. That this dispute concerned agriculture shows both the GOELRO planners' lack of agrarian experience and shared objectives elsewhere and the importance the party attached to the peasantry.

---

new horizons of intervention of the human will and electrotechnology in the elemental packaging of natural forces" according to one GOELRO thesis on rural electrification: TsGANKh f. 3429, op. 1, ed. kh. 1953, 17.

[66] Edmund N. Todd, "Electric Ploughs in Wilhelmine Germany: Failure of an Agricultural System," *Social Studies of Science* 22 (May 1992): 263–82.

[67] A considerable literature on Lenin and the electric plow exists. See, e.g., P. P. Kovalev and A. A. Novikova, "V. I. Lenin i sozdanie elektroplugov (1920–1922 gg.)," *Istoricheskii arkhiv*, 1956, no. 4: 3–38.

[68] TsGANKh f. 5208, op. 1 (2), ed. kh. 1, 45–45b.

[69] Ibid., 47–48. TsGANKh f. 3429, op. 1, ed. kh. 1953, 19.

[70] Aktsionernoe obshchestvo po elektrifikatsii selskogo khoziaistva, *Elektroselstroi i ego deiatelnost* (Moscow: Novaia derevnia, 1924), 30.

[71] Naum Jasny, *Soviet Economists of the Twenties: Names to Be Remembered* (Cambridge: Cambridge University Press, 1972), 31, 196.

[72] 26 June, *Trudy GOELRO*, 160–65; 13 July, ibid., 166.

[73] TsGANKh f. 5208, op. 1 (2), ed. kh. 1, 92–93. For Ballod, see ibid., 6, 15, 17, 22, 24, 52, 73–75.

In June, Mikhail Shatelen suggested the alternative of small village stations to provide light and power for milling and butter churning. Domestic industries would provide the equipment, standardized for ease of production, repair, and cost.[74] Shatelen claimed that inadequate data did not allow him to conclude whether small stations could electrify the countryside more easily and quickly than regional stations, a politically safe conclusion.[75] Shatelen offered a different technical approach to a political question of how the countryside could best be electrified. GOELRO thinking focused on the regional station; local stations offered radically different implications for the social, political, and economic organization of electrification. GOELRO ignored small stations in 1920, but they resurfaced in 1921 as an alternative to the GOELRO plan.

Lenin's dream of electrification to bring the peasant the culture of the city dweller was a revolutionary concept, but it excited neither the agricultural establishment, which had more immediate priorities, nor electrical engineers, who faced more interesting challenges. Whereas other sectors, notably industry, actively clamored for electrification, it is not surprising that agriculture ended up high in theoretical promise and low in actual achievement.

GOELRO, the Document

IN DECEMBER 1920, GOELRO presented the 8th Congress of Soviets with an impressive document of more than five hundred pages.[76] The report consisted of separate plans by region and economic sector with demand forecast to 1930, in keeping with the plan's expected ten-year duration.[77] The basic premise was simple: "To compose a project for the electrification of Russia—this means providing outstanding leadership for all constructive economic activities [and] building the basic

---

[74] TsGANKh f. 5208, op. 1, ed. kh. 53, 2–10.

[75] TsGANKh f. 5208, op. 1 (2), ed. kh. 1, 81b.

[76] A microfilm of the original exists at the Library of Congress but could not be found by the staff. This section is based on the reprinted material in *Plan elektrifikatsii GOELRO* (Moscow: Politizdat, 1955), and *50 let*. The quantitative data are internally inconsistent (e.g., *Plan GOELRO*, 352). Among the possible causes are the evolving sizes of stations, confusion between produced and available output, and poor mathematics. I have found no cause to doubt the veracity of these documents.

[77] Krzhizhanovskii, *Ob elektrifikatsii (rech na 8-m sezde Sovetov)* (Moscow: Gosudarstvennoe izdatelstvo, 1921), 29. Absent is an explicit rationale for the ten-year period. In all likelihood, a decade was sufficiently long to accomplish the plan and sufficiently simple for people to comprehend.

scaffolding for the realization of a single state plan for the economy."[78]
GOELRO promised that electrification would accelerate economic reconstruction while simultaneously transforming the country from a
poor cousin of Western Europe into a modern, cultured society saturated with electric light and radios.

The commission's rationale for a state economic plan confounded
revolutionary and economically pragmatic arguments. The report's introduction stated: "We are working not only for ourselves and our
contemporaries but for the workers of all the world and for their coming destiny."[79] Except for agriculture, the technical sections avoided
this revolutionary tone. The GOELRO planners viewed Germany and
other capitalist countries as sources of high technology, funding, and
ideas, not as potential targets for revolution. This initial emphasis on
international revolution represented both a belief in the future development of socialist revolutions abroad and an attempt to garner political support from those Bolsheviks pressing for revolution in Europe
rather than immediate reconstruction in Russia. Could example, not
armed revolt, bring the West to socialism? In 1917, most socialists,
including Lenin, saw the October revolution as the first of many revolutions; by late 1920, elite opinion—aided by the failure of socialist
revolutions in Europe and wide-scale domestic destruction—had partially swung to the concept of developing Soviet Russia first.[80] By emphasizing both reconstruction and revolution, the commission ap-
pealed to both those who wanted to rebuild Soviet Russia first and
those who wanted to spread socialism abroad.

After these sweeping revolutionary statements, which repeatedly
invoked Ballod, the introduction abruptly enunciated five elementary
rules of administration that sharply contrasted with the prior heady
claims: do not delay what can be done now, do not overestimate your
strength, have a knowledge of productive forces, correctly estimate
the work's difficulty, and finish a project after starting it.[81] The implications were clear: dreams should not proceed faster than the ability
to fulfill them. A sobering reminder of the undertaking ahead, the
inclusion of these rules appears defensive, protecting the planners
from the Scylla of proposals for even more rapid industrialization and
the Charybdis of criticisms of overextension.[82]

The GOELRO report stated that only a rapid increase in labor pro-

[78] "Elektrifikatsiia i plan narodnogo khoziaistva," *50 let,* 88.
[79] Ibid., 89–90.
[80] Stephen F. Cohen, *Bukharin and the Bolshevik Revolution: A Political Biography, 1888–1938* (New York: Vintage, 1973), 55–56, 128–29.
[81] *50 let,* 91.
[82] Ibid., 97–98.

ductivity would overcome the country's economic devastation. The answer was to intensify, mechanize, and rationalize labor, all accomplished best by electrification. Flowing directly from war communism, these precepts promised that Russia's "difficulties can be overcome by the GOELRO project of electrification, and therefore its materials should be read by all interested organs as one single plan of the state economy, *although in its first approximation.*"[83]

Solving the fuel crisis received first priority. GOELRO presented electrification as a means to increase fuel production, a mechanism for transmitting power, a more efficient method of utilizing fuels, a way of reducing demands on railroads, and, ultimately, a method of developing more rationally sited new industries. The short-term unification, repair, and use of existing powerplants would help overcome the fuel, food, and transportation crises.[84] Hydropower occupied pride of place among power sources, an interesting priority for a country with only two small commercial hydroplants. Large dams would supply existing users but, more important, provide a qualitatively new form of power for new industrial centers built near their raw materials. Cheap, abundant "white coal" would advance navigation and irrigation and aid Russia's exports by creating new industries, such as aluminum refining in the north, and by rebuilding and expanding prewar exports, such as southern grain. If necessary, thermal stations would be combined with hydroplants; the former would be built first, then become peak-load plants.[85]

The GOELRO plan demanded top economic priority for the creation of a framework of high-capacity railroads in the Donets, Central Industrial, and Urals regions, based on American and German experience. Electrification would play a crucial role by greatly reducing operating costs, improving efficiency, and lowering fuel consumption. Substations would electrify areas near railroads to develop local industry and culture. Newly developed waterways and ports would complement railroads. Increasing capacity to overcome the transportation crisis was the immediate goal; the improved efficiencies of electrification would come later. A few years of exports would repay the costs of this limited electrification.[86] The report reserved other means of transportation, such as automobiles, for future discussion.

To rationalize industry and close the growing economic gap be-

[83] Ibid., 91–93, 104–5, emphasis added. This caveat received prominent attention throughout the trouble-plagued 1920s; see, e. g., E. Ia. Shulgin, "K peresmotru plana elektrifikatsii," *Planovoe khoziaistvo*, 1925, no. 2: 22–23.

[84] "Elektrifikatsiia i toplivosnabzhenie," *50 let*, 106–61.

[85] Ibid., 158–60; see also the respective regional sections, e. g., 613–14 for Siberia.

[86] Ibid., 147–57, 222–32, 248.

tween Russia and the West, GOELRO recommended a course that sounds quite similar to American precepts of scientific management, precepts that had reached the Russian electrotechnical community before the war.[87] The commission urged the state to simplify, unify, and standardize manufactured products; maximize mechanization; implement a more complete division of labor; and create large centers of production and decrease the number of enterprises. As in the West, electrification would change the nature of Russian industry, by locating industries closer to raw materials and integrating kustar and remeslo handicrafts into the national economy. Among industries, primus inter pares was metallurgy, followed by other extractive and processing industries.[88]

An example of GOELRO's sectoral analyses is its treatment of cement.[89] Before the war, fifty-eight factories produced 27 million barrels annually. Annual per capita consumption hovered at 0.1 barrel, compared with 0.5 in Germany and 1.0 in the United States. GOELRO planned to produce 50 million barrels yearly by 1930 and double per capita consumption. This output would require 100,000 workers, 120 MW of generating capacity, and 1 billion kWh. The major bottleneck would be obtaining the necessary equipment, previously imported. GOELRO intended to resume imports, but it did not rule out the development of domestic manufacturing.

For all sectors of industry, GOELRO estimated that production would increase by a factor of 1.8, the number of workers by 1.2, fuel consumption by 1.4, and power usage by 1.7; that is, mechanization, rationalization, and intensification would increase industrial efficiency.[90] GOELRO's industrial plans assumed strong ties with the world economy. As under tsarism, raw materials would be exported and technologies imported. Now, however, Russians—guided by a comprehensive, long-term plan—and not foreigners would control Russian industry.

The section on agriculture, although twice as long as any other section, lacked the wealth of information elsewhere. An aggressive nationalistic-ideological rhetoric substituted for specifics: *"The depth and scale of influence of our regional electric stations on all our national economy can and must exceed European and American norms."*[91] The future of Soviet agriculture would rest in electrified state farms, operating with

---

[87] A. Pankin, "Nauchnaia organizatsiia truda," *Elektrichestvo*, 1910, no. 10: 297–305; no. 11: 327–35.

[88] *50 let*, 250–57.

[89] Ibid., 263–64.

[90] Ibid., 269–70.

[91] Ibid., 205.

the certainty and precision of industrial plants to produce large yields and fruitful propaganda. GOELRO tightly linked inexpensive electric energy for agriculture to the development of regional stations.[92] Local electrification would occur only after the development of regional stations and large transmission networks. Rural electrification remained a poor country cousin to transportation and industry. It could plan its own development only after its bigger relatives determined the geographic and temporal framework of regional stations.

GOELRO proposed the construction of 112 regional stations with three levels of priority.[93] The thirty highest priority stations would add 1,750 MW, approximately five times total prewar capacity.[94] Together with the 5,916 MW in the remaining eighty-two stations, GOELRO would give Russia more capacity than all American utilities in 1912, an astounding jump.[95] In developed areas, electrification would serve primarily as a method of fuel substitution for existing industries. In most of the country, electrification and the concomitant industrialization would lift entire areas into the economy and culture of modern Russia. Priority, as Table 6.2 shows, went to the industrialized areas, not to the industrializing: the immediate emphasis was on European Russia, especially Moscow and Petrograd.

The more developed Central Industrial and Northern regions would receive a third of the thirty first-priority stations and 555 MW, with an average station capacity of 55 MW. The Southern region accounted for five stations and 560 MW, with an impressive 82-MW average for the four thermal stations and a stunning 230 MW for the Dniepr hydroplant. These five stations would electrify the Donets coal region and its heavily used railroads, which transported Ukrainian grain to Odessa for export. The remaining four regions accounted for half the stations but only 635 MW, for an average of 42 MW per station. The geographic distribution of the eight-two second- and third-priority stations also favored the Central Industrial and Northern regions, followed by Siberia. Larger than most of the prewar German and all tsarist utilities, these stations constituted a large technological leap forward, a leap for which Russia's electrotechnological manufacturing was ill prepared.[96]

The Northern Region's priority was autarkic industrial develop-

[92] Ibid., 179, 216.
[93] *Plan GOELRO*, 264–67, 352, 424, 497, 535–37, 590, 607–8, 649.
[94] "Biulleten," *Elektrichestvo*, 1927, no. 1: 43.
[95] "Statistics on the Operations of the Electric Light and Power Industry," *Electrical World*, 7 January 1928, 32.
[96] "German Central-Station Statistics," *Electrical World*, 10 January 1914, 105–6.

Table 6.2. Regional stations by geography and priority

Region	1st priority[a]	2d and 3d priority[a]	Total[a]
Northern	4/195 (11)	35/1266 (21)	39/1461 (19)
Central	6/360 (20)	10/960 (16)	16/1320 (17)
Industrial			
Southern	5/560 (32)	3/410 (7)	8/970 (13)
Urals	4/210 (12)	5/395 (7)	9/605 (8)
Volga	4/120 (7)	5/240 (4)	9/360 (5)
Caucasus	4/155 (9)	8/328 (6)	12/483 (6)
Siberia	3/150 (9)	16/2317 (39)	19/2467 (32)
and Turkestan	30/1750	82/5916	112/7666

Source: *Plan elektrifikatsii GOELRO* (Moscow, 1955), 264–67, 352, 424, 497, 535–37, 590, 607–8, 649.

[a] Number of stations/installed capacity (MW). Percentage of total capacity in parentheses.

ment. Never again would Petrograd depend on foreign energy. Future industrialization would occur closer to raw materials, a major spatial shift possible only because of hydropower. Yet, even with the transfer of industries from the former capital, the seaport's estimated nonindustrial consumption in 1930–32 demanded new thermal stations as well as hydropower. The need to export dictated the development of conduits to the West, notably by White Sea and Baltic Sea ports connected to the interior by railroads. Wood and metals, including aluminum, would be the main exports.[97]

As the focus shifted from the general to the specific, the first bloom of peat and hydropower faded. In principle, peat-fueled stations for Petrograd were an excellent idea, but they demanded a tremendous investment in extraction, processing, and transportation. Hydropower demanded greater capital investment and longer time to build than thermal stations, as the Volkhov and Svir projects annoyingly demonstrated. GOELRO envisioned the construction of thermal stations only for the short-term baseload. Once hydropower flowed, these thermal stations would become peak-load plants.[98]

Substitution for southern oil and coal drove GOELRO's plans for the Central Industrial Region. Industry would satisfy domestic before foreign demand, a reversal of the Northern Region's policy but appropriate for Moscow's traditional domestic orientation. As in the north,

[97] *Plan GOELRO*, 231, 236, 248, 250–57.
[98] Ibid., 234.

"it will be indispensable to resort to technical and material assistance from abroad, though due to the poverty of our country this will be possible only in a limited manner."[99]

Outside the Northern and Central Industrial regions, GOELRO's estimates became more approximate. Resources and geography needed exploration, data contained major lacunae, and little industrialization had occurred. Estimates of future consumption represented more anticipation than analysis. Nonetheless, GOELRO's work represented a major step in contemplating the entire country's economic development. Plans for the Southern Region, for example, envisioned the creation of a vast industrial base, consuming 1,000 MW, with other uses, including railroads, consuming an equal amount. But the first steps were small: unification of existing powerplants and construction of a 10-MW anthracite station at Shterov. Basic questions remained unanswered, even if well framed. Should new electrical energy go to new or to existing coal mines? Should the Dniepr hydroelectric station at Aleksandrovsk have one dam or two? What were the appropriate economics of scale for development?[100]

The least developed and most difficult area of GOELRO's calculations was financing. The commission estimated ten-year expenditures of 18 billion gold rubles, of which only 1.2 billion rubles (7 percent) would go directly to electrification. Transportation and industrial development would consume most of the remainder. Exports of grain and raw materials would fund two-thirds of these expenditures, leaving a deficit of 6–7 billion rubles to be filled by foreign loans and concessions.[101] GOELRO also calculated the materials needed to construct the thirty first-priority stations. Ranging from 150 million bricks to 450,000 square meters of boiler surface, these data indicate the vast magnitude of the anticipated physical construction.[102]

Built on inadequate information, admittedly vulnerable to many national and international factors over which it had no control, and fraught with internal contradictions, the GOELRO plan was nonetheless ambitious. The 7,666 MW from all 112 regional stations represented a massive twentyfold increase from prewar capacity. The technical requirements boggled the Russian mind of 1920: creation of extensive high-voltage transmission networks; doubling, tripling,

---

[99] Ibid., 319, 371.
[100] Ibid., 442–44.
[101] 50 let, 269–70.
[102] See Plan GOELRO, 212, for the complete list, which refers only to the material needed for GOELRO construction and not for the economic development of the entire nation.

quadrupling of the size of thermal plants; development of boilers to use the variegated local fuels; and hydropower. Add to this the antici- pated political and social transformations, and one can understand why critics called the GOELRO plan "electrofiction" (*elektrofiktsiia*).[103]

The 8th Congress of Soviets

THE LONG TIME GOELRO needed to produce its plan contrasted sharply with the hurried, almost frenzied efforts to present it, efforts that resembled "storming" to fulfill quotas, a practice that started in 1920 and reached maturity in the Stalinist *udarnichestvo* shock move- ment.[104] The rushed preparations for the February central executive committee meeting can be excused as a last-minute development. Yet the same frantic haste marked the 8th Congress of Soviets, eight months after the original deadline for GOELRO's report. Although GOELRO had prepared since October for its public unveiling, Lenin had to write to include the GOELRO presentation at the 22–29 De- cember congress. He also had to ensure the installation of a map at the Bolshoi Theater and the printing of 5,000 copies of the fifty-page synopsis, which did not reach the VSNKh printers until 29 Novem- ber.[105] In a telling commentary on Moscow's economic straits, only promises of food and materials plus Lenin's participation in "long ne- gotiations, arguments, entreaties" ensured prompt printing.[106]

The delegates received an elaborate show—copies of the GOELRO report, a gigantic map portraying the electrified Russia of ten years hence, and speeches by Lenin and Krzhizhanovskii. The effort pub- licized electrification and overwhelmed doubters and opponents. The congress met in the "cold, weakly lighted theater," drowning in shadows as "brilliant light flooded only the stage of the theater and the gaze of all the delegates remained on the large map," according to one delegate.[107] A more critical participant compared the map with

[103] *Elektrifikatsiia Rossii: Trudy 8-ogo Vserossiiskogo elektrotekhnicheskogo sezda v Moskve 1–10 oktiabria 1921* [8th VES] (Moscow: Gosizdat, 1921), 1: 5.

[104] Remington, *Building Socialism*, 156; Hiroaki Kuromiya, *Stalin's Industrial Revolution: Politics and Workers, 1928–1932* (Cambridge: Cambridge University Press, 1988), 115–28.

[105] 19 October, *Trudy GOELRO*, 177–78; Lenin, *Collected Works*, 45: 63; Steklov, *Lenin i elektrifikatsiia*, 59.

[106] Vladimir I. Aleksandrov, "Po zadaniiu Ilycha," *SRE*, 137–39.

[107] V. A. Smolianinov, *SRE*, 59.

holiday illumination.[108] Moscow's generating capacity was so low that the congress's consumption produced blackouts elsewhere.[109]

The 8th Congress of Soviets formally merged the utopia of the electrical engineers, a national program for electrification, with that of the Bolsheviks, a path to communism. When Lenin declared "Communism is Soviet power plus the electrification of the whole country," he linked the prestige and future of the Communist party with the goals of a group of mostly nonparty engineers and specialists.[110] The theoretical cement for this union was Lenin's justification of electrification as the transformational path for Soviet Russia. For Lenin, electrification promised a new era when politicians would speak less and engineers and agronomists more. Electrification might seem utopian, but it was necessary: "Only when the country has been electrified, and industry, agriculture, and transport placed on the technical basis of modern large-scale industry, only then shall we be fully victorious" over capitalism at home and abroad.[111]

Lenin convinced the delegates that electrification and its accompanying economic plan would achieve the first program of the party—its political program—not just in Russia but worldwide. More practically, "without a plan of electrification, we cannot undertake any real construction. We must adopt a definite plan . . . which will stand before all Russia as a great economic plan designed for a period of not less than ten years, and indicating how Russia will be placed on the real economic basis required for communism." For the party, electrification promised victory over capitalism. For industry, transport, and agriculture, the promise of coordinated reconstruction and growth beckoned. Finally, to win over the potentially most dangerous foes of socialism, the peasants, "we must now try to convert every electric power station we build into a stronghold of enlightenment to be used to make the masses electricity-conscious, so to speak."[112]

Krzhizhanovskii declared that electrification "can be rendered into an active weapon in our hands." At home, the troika of intensification, mechanization, and rationalization of labor would solve the crises in transportation, fuel, food, and labor productivity. Abroad

[108] F. Dan, Dva goda skitanii (1919–1921) (Berlin: H. S. Herman, 1922), 96.

[109] Roger W. Pethybridge, The Social Prelude to Stalinism (London: Macmillan, 1974), 35.

[110] Vosmoi Vserossiiskii sezd sovetov: Stenograficheskii otchet (Moscow: Gosudarstvennoe izdatelstvo, 1921), 30.

[111] Ibid., 28–29, 516.

[112] Ibid., 31, 32, 29.

Electrification and counterrevolution. Courtesy of the Hoover Institution.

"stand opponents, equipped with all the attributes of the strongly developed capitalist economy. It is perfectly clear that in the economic struggle we must be armed in the same areas they are. . . . if we do not work from a base of electrification, our position will be extremely disadvantageous."[113]

A decade of the GOELRO plan would not only heal the wounds of war but nearly double industrial production from the 1913 level. Krzhizhanovskii sounded cautionary only about the number of stations (twenty-seven were more realistic than 112), organizational difficulties, and the enormous investment demanded for hydropower.[114] This change from all 112 stations to the twenty-seven or thirty first-priority stations demonstrated that GOELRO had already tempered its ambitions to political and economic realities.

The congress did not discuss the GOELRO plan; instead, Politburo

[113] Krzhizhanovskii, *Ob elektrifikatsii*, 5–9.
[114] Ibid., 26, 30, 39.

member Lev B. Kamenev created a group headed by Krzhizhanovskii to write a preliminary resolution.[115] The few comments in the stenographic report mostly praised electrification. Grigorii E. Zinoviev, the ruler of Petrograd, hailed Krzhizhanovskii's talk as "the most significant moment of our Congress, when we were all enchanted listeners of the speech of the engineer, who drew for us the future economic development of the country." Leon Trotsky, basking in acclaim for his military leadership and his bold plans for the Commissariat of Transportation, hailed Lenin's identity of socialism as electrification, albeit within the framework of a reconstituted transportation system. In closing the congress, Kamenev declared, "We need to place the political power of the workers on a new technical base. . . . the electrification of Russia [will serve as] the base for Soviet proletariat power . . . to open the road to a new conquest by technology, which will mark the greatest gigantic victory by mankind over the elemental forces of nature."[116]

Only Aleksei I. Rykov and Fedor I. Dan offered critical commentary. VSNKh chairman Rykov agreed on the need for a single plan to organize the economy, but he called the GOELRO plan only preliminary because it had not been approved by the VSNKh presidium. The cautious Rykov remained cool to the GOELRO plan, which threatened VSNKh's economic responsibilities.[117] In critiquing Bolshevik economic policy, Dan, one of two token Mensheviks, disparaged Lenin's cult of the engineer and agronomist, insisting that political power, not electrification, was the issue. The Menshevik declaration supported a "general, single plan of economic life for the country," but one in which the workers participated and which defended their interest.[118] The GOELRO plan, prepared by a small group of specialists, did not meet those criteria. Dan's comments, unsurprisingly, were ignored. In 1922 from foreign exile he disparaged Lenin's "sufficient number of fools [who believed] this balderdash."[119]

[115] *Vosmoi Vserossiiskii*, 88.

[116] Ibid., 207, 288–89, 258. The dream was shared. Compare Kamenev with Segal describing energy in American utopias: "Wind, water and other natural resources will be harnessed—into electricity above all, its supreme cleanliness and quiet befitting its supreme power"; in Segal, *Technological Utopianism*, 24.

[117] *Vosmoi Vserossiiskii*, 115–16, 89; D. Baevskii, "Leninskii plan sotsialisticheskogo preobrazovaniia Rossii i GOELRO," *Voprosy istorii*, 1947, no. 3: 9.

[118] *Vosmoi Vserossiiskii*, 34–35, 51. See also Vera Broido, *Lenin and the Mensheviks: The Persecution of Socialists under Bolshevism* (Boulder: Westview, 1987), 83–85, and Leopold H. Haimson, ed., *The Mensheviks: From the Revolution of 1917 to the Second World War* (Chicago: University of Chicago Press, 1974), 236–37, 247.

[119] Dan, *Dva goda skitanii*, 96.

The congress called the GOELRO plan the "first step of a great economic undertaking." It ordered the appropriate state organs to complete and implement the plan, including "the widest possible propaganda."[120] Electrification was now official state policy. Its long journey to implementation had, however, only started.

Politics

ALTHOUGH THE 8th Congress of Soviets approved the GOELRO plan, discussion continued, as did debate about the future of planning and a single economic plan.[121] Making electrification a state technology did not have universal party approval. Planning enthusiasts, such as Vladimir P. Miliutin, the VSNKh deputy chairman, gave transportation, fuel, and industry higher priority than electrification.[122] These discussions over the GOELRO plan and the future of planning foreshadowed the industrialization and superindustrialization debates of the mid- and late 1920s.

A year passed between the GOELRO plan's acceptance by the 8th Congress of Soviets and approval by the 8th All-Russian Electrotechnical Congress and the 9th Congress of Soviets. One important political and technical change was the greater attention to small rural stations. Their advocates were more interested in the political and social consequences of rural electrification than the technocratic efficiency of regional stations. Although interest in rapid rural electrification grew through the mid-1920s, this decentralized approach never captured the allegiance of the party or the electrical engineering leadership, both of whose interests favored centralized control and development. The debates about the GOELRO plan revolved around whether a decentralized or centralized approach would best electrify Russia. Rarely, however, were debates framed in these terms. Instead, advocates and, to a lesser degree, critics used more overtly political language.

The GOELRO plan received "assaults both from the right and the

---

[120] *Vosmoi Vserossiiskii*, 271–72.

[121] Carr, *Bolshevik Revolution, 1917–1923*, 2: 370–78; A. D. Pedosov, "Ideinaia borba vokrug plana GOELRO i zashchita ego V. I. Leninom," *Voprosy istorii KPSS*, 1969, no. 7: 26–37.

[122] E. g., M. Markovich, "Edinyi khoziaistvennyi plan," *Narodnoe khoziaistvo*, 1920, nos. 15–16: 2–6; V. P. Miliutin, "O metodakh razrabotki edinogo khoziaistvennogo plana," *Ekonomicheskaia zhizn*, 19 February 1921, 1; L. N. Kritsman, *Novaia ekonomicheskaia politika i planovoe raspredelenie* (Moscow: Gosudarstvennoe izdatelstvo, 1922); Lenin, *Collected Works*, 32: 137–45.

left," according to Krzhizhanovskii in October 1921.[123] The right attacked the GOELRO plan as too ambitious, and the left attacked it as insufficiently ambitious. Krzhizhanovskii's confounding of technical and political criticism established a framework for viewing any criticism as political opposition: through the decades, conservative doubters about GOELRO's technical feasibility and Trotskyists have received the majority of official opprobrium.[124] Labeling critics as left and right opposition split the ranks of critics and equated loyalty to GOELRO with loyalty to the Communist party, an effective tactic to minimize dissent.

One significant line of criticism attacked the plan as a conservative continuum instead of a foundation for a revolutionary transformation of society. These criticisms reflected opposition to the transition from the command economy of war communism to the more market-oriented NEP as much as strictly technological concerns, and they threatened to undermine support for GOELRO. This critique judged GOELRO insufficiently revolutionary, but Krzhizhanovskii castigated its denial of economic reality: "You cannot compare us with the United States, for we are an especially victimized country [*postradavshaya strana*]. . . . We have the greatest possibilities but in acting on them we must be extremely cautious."[125]

This leftist critique contained two themes, an assumption that heroic efforts by Bolsheviks could overcome any difficulty and a belief in the need for a third revolution to consummate the two revolutions of 1917. Technological naivete and organized revolutionary enthusiasm, reasonable attributes for the triumphant Red Army, marked this approach. Such attributes did not, however, encourage alliances with the engineers who advocated local electrification, because the latter came from a different, non-Bolshevik background and viewed the issue as how to electrify the countryside most quickly, not how to continue the revolution.

The faith in heroic efforts, based on war communism and a feature of Stalin's superindustrialization, was best embodied in Ia. M. Shatunovskii's 1921 pamphlet *Belyi ugol* (White Coal), which urged the

---

[123] *8th VES*, 1: 31.

[124] E. g., I. A. Gladkov, *V. I. Lenin i plan elektrifikatsii Rossii* (Moscow: Gosplanizdat, 1947), 73; V. Iu. Steklov, *Elektrifikatsiia strany sovetov* (Moscow: Partizdat, 1936), 120–30. The preface to the 1955 reprint of GOELRO colorfully described the plan's opponents as "enemies of socialism, left and right restorers of capitalism, opportunists of all colors, different yes-men bearing malice to Soviet power, imperialists" and, worse, Trotskyists; *Plan GOELRO*, 16, 19.

[125] *8th VES*, 1: 31–32.

mobilization of labor to build hydroelectric plants for Petrograd within a year.[126] For Trotsky, who defended the pamphlet against Lenin's criticism, the basic problem was that Petrograd in its present form was not viable. The alternatives were to shut the city down or electrify it on a priority basis.[127] A year earlier, the Council for the Economy for the Northern Region had developed a more subdued plan to build the Volkhov and Svir hydrostations in three to four years to electrify "Red Peter."[128] In his *Revoliutsiia i elektrifikatsiia* (Revolution and Electrification), Bolshevik Boris Kushner hailed the "permanent revolution" in electrotechnology and urged pressing the third, social revolution in the countryside by immediately building hundreds of small stations instead of waiting a decade for regional stations. Kushner approved the GOELRO plan but thought that revolution in the countryside demanded action within months, not years.[129] Only small stations could answer that demand.

The label of the so-called rightist criticism covered many who objected to parts of the GOELRO plan. Their concern was the plan's practicality and specifics in an economically devastated Russia. They viewed the future with more caution—or realism—downgrading the promise of electrification, criticizing its optimistic assumptions, and approaching it as the basis for a plan, not the plan itself.[130] The Menshevik Dan expressed such doubts when he noted that, although economic reconstruction was necessary, "electrification itself demands material, technical, social, and cultural prerequisites, which contemporary Russia does not have."[131] But Dan wrote outside Russia and out of the action.

In February 1921, the government ordered the convocation of the 8th All-Russian Electrotechnical Congress to discuss the technical and economic questions of implementation and rebuke those "skeptics and whisperers who have maliciously mixed the term electrification with the term 'electrofiction'."[132] Though mandated for April, the congress did not convene until October 1921. The reason for the delay is

[126] Leon Smolinski, "Planning without Theory," *Survey*, July 1967: 116, 119.

[127] For the exchange of letters, see *The Trotsky Papers*, ed. Jan M. Meijer (The Hague: Mouton, 1971), 2: 446–51; Lenin, *Collected Works*, 45: 137; *Leninskii sbornik*, 20: 208–9.

[128] M. Grandov, "Ne otstavaite ot Pitra," *Bednota*, 24 April 1920, 1.

[129] Kushner, *Revoliutsiia i elektrifikatsiia*, 10, 13, 22–25, 27.

[130] M. Savelev, "K itogam VIII-go sezda Sovetov," *Narodnoe khoziaistvo*, 1920, nos. 19–20: 19–24; Steklov, *Lenin i elektrifikatsiia*, 291–93.

[131] Dan, *Dva goda skitanii*, 96.

[132] "Dekret SNK o sozyve VIII VES," 8 February 1921, in *K istorii plana elektrifikatsii sovetskoi strany*, ed. Iuri A. Gladkov (Moscow: Gospolitizdat, 1952), 217–18; *8th VES*, 1: 5. This statement was part of the official introduction.

not fully clear, although, if the 8th Congress of Soviets was any indicator, organizational shortcomings and the other obligations of GOELRO's personnel contributed to the delay.

Compared with prewar electrotechnical congresses, the 8th did well, drawing 893 members from 102 cities and 475 guests. More than 500 participants came from Moscow and another 150 from Petrograd, indicative of a shift from the old to the new capital, although difficulties with travel and housing probably limited attendance from outside the capital. Krzhizhanovskii served as president. The congress elected Lenin honorary president, but he did not actually participate.[133] The overall tone of the congress proved more subdued and conservative than that of the 8th Congress of Soviets a year earlier, because the end of the civil war had not resolved the various domestic crises—new problems actually emerged—and, more important, the audience was different. The congress comprised those who would implement the plan and who possessed firsthand knowledge of the task. Although they endorsed the GOELRO plan, the delegates proved anything but a rubber stamp.

The explicit goal of the congress was to approve the plan, but the implicit goal was to rejuvenate the Russian electrotechnical community. Speakers repeatedly praised prerevolutionary links, and the electrical engineers unsuccessfully tried again to establish a high-level state organ for electrification.[134]

The electrical engineering leadership continued to take its ideas, as well as equipment, from the West.[135] The civil war had restricted Soviet access to foreign technical literature, but the resumption of international communications allowed Krzhizhanovksii to boast that, "now we can refer in defense of our position to a whole series of first-class West European authorities" and publications that "graphically show that we were right when we established [GOELRO, for] all progress in worldwide technology is tightly linked to electrification."[136]

Controversy existed about the best path to electrification, but neither about the priority nor the need. The major dispute developed over centralized versus decentralized small- and medium-scale electrification. GOELRO viewed regional stations as a politically necessary

[133] *8th VES*, 1: 4, 14.

[134] P. S. Osadchii, "Organizatsionnye zadachi provedeniia v zhizn plan elektrifikatsii Rossii," in ibid., 1: 142–47.

[135] E. g., reports often began by noting "here it is necessary in advance to say that the equipment must be received from abroad," as in Krug's speech on the electrotechnical industry; ibid., 1: 69.

[136] Ibid., 1: 28.

step in the advancement and transformation of Russian society. Advocates of smaller stations considered a decentralized approach necessary due to the state of existing Russian technology and economics. Other debates concerned inadequate information and the fit between the GOELRO estimates and local reality. Most of these debates were responses to GOELRO plans, while the dispute over scale grew from the change in political environment from 1920 to 1921. The GOELRO planners acknowledged the need for objective criticism and emphasized working with local experts because "we cannot blindly believe in paper and statistics."[137]

Several projects received sharp criticism. Particularly devastating was the rector of the Saratov Polytechnical Institute, Lagovskii, who described the serious flaws in the three stations planned for his region, flaws based on inadequate information and overly optimistic assumptions. He concluded to applause that, if his region proved typical, "I can boldly say that this project is wholly insufficiently based, and that to hail it as the implementation of the idea of electrification is impossible and it would be impossible to stamp on it the hope that it could be brought to life."[138] Other speakers described similar problems in their geographic and industrial areas. The charge of timidity could be raised against them, but they dealt on a microlevel with the issues that GOELRO handled on a macrolevel. From their perspective, the GOELRO plans seemed confirmation that "lovers of fantasy are everywhere," that the plans had little connection to reality and were far too optimistic. In response, Krzhizhanovskii said, "We will make errors and omissions but better than the mistake of doing nothing."[139]

Doubts also surfaced about GOELRO's financial and export assumptions. A. F. Levitskii of the Petrograd Technological Institute claimed that the GOELRO plan would cost 60–70 billion contemporary gold rubles, four times more than the estimate in prewar rubles. Furthermore, GOELRO excluded the costs of industrial reconstruction and a transmission network, which added another 100 billion gold rubles. Finally, the export estimates seemed overly optimistic.[140] In replying to such "slanderers," Krzhizhanovskii agreed that prewar

[137] Osadchii, "Organizatsionnye zadachi," 142; ibid., 1: 33.

[138] Ibid., 1: 96.

[139] Ibid., 1: 84, 91. Vigura, a VSNKh representative, also thought twenty years a more likely period, derided village electrification, and concluded by quoting an English saying, "Don't bite off more than you can chew."

[140] Ibid., 1: 75–77. Some GOELRO regional analyses did include the cost of transmission networks. For doubts about exporting a billion rubles of timber, see Artemev of the Petrovskii Agriculture Academy; ibid., 1: 82.

prices were misleading but argued that fluctuations in the ruble and the blockade-imposed lack of information had made any alternative impractical. As for exports, the estimates were feasible, especially for oil. In an interesting volte-face, the previously favored Grinevetskii now belonged "in the category of our opponents, for he has become a simple apologist of large capitalist industry."[141] This switch came from an ideological reassessment of Grinevetskii's anti-Communist tone and not from any actions of the engineer, who had died two years earlier.[142]

These debates paled into insignificance compared with those on the scale of electrification. The issue boiled down to resources and decision making. Should electrification be directed from above or arise democratically from below?[143] Was local electrification "economically illiterate," wasteful, and, as viewed in the West, neither economically nor technically rational? Or were small stations the best means to electrify rural regions?[144]

The GOELRO plan concentrated investment in regional stations to maximize return from its limited resources and create the "first socialist accumulation." Small-scale electrification was a minor local matter to be aided by limited state support.[145] By contrast, advocates claimed that small-scale electrification could transform the countryside years before GOELRO's industrially oriented regional stations reached rural areas. Instead of a web of transmission lines radiating within a few economically developed regions, thousands of small-scale stations would saturate the country, producing a very different economic and political map from GOELRO's. Unless it provided this broad-based electrification, the GOELRO plan would follow the old foreign-dominated tsarist path of large-scale centralized development directed by the state.[146]

Small-station advocates demonstrated the technological viability of their approach. In a continuation of his 1920 GOELRO paper, Shatelen and B. E. Borovev proposed standardized 3-phase AC stations in

[141] Ibid., 1: 87–88.

[142] Gregory Guroff, "The State and Industry in Russian Economic Thought, 1909–1914" (Ph.D. diss., Princeton University, 1970), 267.

[143] Introduction to the technical-economic section, 8th VES, 2: 5.

[144] See Krzhizhanovskii, ibid., 1: 89; Kozmin, ibid., 1: 79; M. A. Shatelen and B. E. Borovev, "Snabzhenie selsko-khoziaistvennykh raionov elektricheskoi energiei ot mestnykh stantsii maloi moshchnosti," ibid., 2: 53–54.

[145] Ibid., 1: 89–90.

[146] F. K. Ryndin, "O vozmozhnosti ustroistva elektricheskogo osveshcheniia v derevne v sviazi s sozdaniem semi raionnykh melnits," ibid., 2: 24–34; Kozmin, ibid., 1: 79.

five sizes from 17 to 100 kW. When the grid from regional stations eventually reached these stations, they could connect without problems. Unlike regional stations that depended on key foreign technologies, Russian manufacturers would equip small stations, thus saving valuable foreign exchange and reviving local and domestic industries. One proposal envisioned milling concessions with payment in grain in exchange for village electrification.[147] Lacking political ties, the advocates of small-scale electrification carried the day in ingenuity but lost the war.

As is not surprising, state-dominated electrification and industrialization remained the top priority. Despite doubts about undertaking too much, the congress resolved that the GOELRO plan "on the whole is the correct scheme by which to construct a state-planned economy," although actual fulfillment depended on domestic and international conditions. GOELRO ceded some ground on "*the necessity of the planned cultivation and implementation*" of small and medium-size stations, which the congress viewed as worthy of state sponsorship.[148]

Three factors, two external to GOELRO, explain the increase in interest in decentralized, small-scale electrification. Most important was the change in the political environment. GOELRO originated in the centralized, militarized atmosphere of war communism. The electrotechnical congress occurred in the initial surge of decentralization under the NEP which emphasized closer urban–rural links. The struggle between cities and the ETO to control urban utilities (discussed in Chapter 7) further increased the political opposition to centralized control of electrification. The internal factor concerned the plan itself. The huge costs of the GOELRO plan, combined with its known errors, concerned many. More worrisome was the realization that GOELRO would not reach large areas of the country for over a decade. If bridging the town–country gap meant electrifying rural areas, then a faster approach was required. Thus, a combination of pressures against centralized control and the realization of deficiencies contributed to the reaction within the electrical engineering community against the GOELRO plan.

Criticism and alternatives to the plan developed after its unveiling.

[147] Shatelen and Borovev, "Snabzhenie selsko-khoziaistvennykh raionov," 52–71; E. Liskun, "Selskoe khoziaistvo severnoi oblasti v sviazi s planami elektrifikatsii," ibid., 2: 72–86; Ryndin, "O vozmozhnosti," 24–34.

[148] Ibid., 1: 163; quote from P. Kozmin, "K itogam VIII-ogo Vseross. elektrotekhn. sezda," *Narodnoe khoziaistvo*, 1921, nos. 11–12: 99.

That they failed to overturn the centralized plan indicates not the superiority of GOELRO's technological vision but the weakness of its opponents. Despite revealing political, economic, and technical flaws and an environment that increasingly favored a more decentralized economy, critics proved unable to form effective networks to oppose GOELRO and provide politically and technically viable alternatives. Krzhizhanovskii and GOELRO had linked the electrical engineering and political leadership. Their critics did not.

The 8th Congress of Soviets gave the GOELRO plan an essential political mandate. The 8th All-Russian Electrotechnical Congress gave it technical approval but veered to the side of caution. For one supporter, the model was the United States's emergence from a ruinous civil war to host the world's most impressive technical exhibition only a decade later.[149] The problem with this analogy was that the American north's industrial strength remained untouched by the horrors of war. Even the application of a foreign analogy demanded adjustment to local context.

The Foreign Role

THE MASSIVE ROLE of the West as a supplier—of legitimation, data, ideas, equipment, and financing—could, but should not, be viewed primarily as evidence of Russian dependence. Rather, the Western role shows the extent to which Russian electrical engineers considered themselves part of the international electrotechnical community and how far they had to go to catch up.[150]

Several lines of Western involvement ran deeply throughout the GOELRO plan and the accompanying discussions. Foremost was Western progress in electrical engineering, which provided inspiration and argument for Russian engineers: "If they can do it, so can we—and since they are doing it, we should." Russian electrical and political leaders expected Western technology and capital to play a large role in implementing the GOELRO plan in return for Russian foodstuffs and raw materials.

Extremely conscious of the need to pay for this technology, GOELRO emphasized exports. In agriculture, Ugrimov cited wheat

[149] Kozmin, "K itogam," 105.
[150] GOELRO frequently cited European statistics to demonstrate Russian inferiority, such as comparisons of fuel use; see *50 let*, 95.

exports as justification for feeding Soviet citizens at a level just above hunger. The Gorev–Shvartz plan for northern electrification envisaged the development of an export-oriented aluminum industry accompanied by the transformation of Murmansk into a deep sea port. The Dniepr development envisioned Aleksandrovsk as a second Odessa. For Siberia, Krzhizhanovskii declared, "it is vital to display before Europe the possibility of using the riches of west Siberia, especially in agriculture, for an escape from the forthcoming world food crisis."[151]

This macroeconomic emphasis typified GOELRO's wide scope and concern with material balances. In contrast to the claims of the preface to the 1955 reprint that "independence and defensive capacity of the Soviet state were the central, leading goals" of electrification, GOELRO assumed a productive economic interdependence between the capitalist countries and the world's first socialist state.[152] Underlying this wide-ranging plan was the assumption of normal international economic relations. GOELRO might publicly espouse the export of revolution and international proletarian solidarity, but conventional trade dealings would be the order of the day. This presumption—astounding in 1920 when the young state was just reestablishing trade links and remained diplomatically unrecognized by the major powers—owed much to the engineering worldview. The GOELRO engineers had lived in the West, worked for foreign firms, and followed Western developments. They wholeheartedly advocated resuming the import of advanced technologies. The question was not "if" but "how much." To electrify properly required extensive dealings with the West.

Under GOELRO's plan, Soviet Russia would struggle not to build socialism autarkically in one country or to export the dictatorship of the proletariat but to resume international relations, albeit on its own terms. The Russians very much wanted to return to the international electrotechnical community, where they correctly thought they belonged. GOELRO was not just a plan for reconstructing Russia but also a statement that Russian electrical engineers were members of a larger engineering fraternity as well as the technocratic modernizers for the new state. It was a statement echoed by other Russian scientists and engineers.[153]

---

[151] "Protokol," 24 April, *Trudy GOELRO*, 128–29; 22 May; 26 October, ibid., 179.

[152] "Velikii ekonomicheskii plan," *Plan GOELRO*, 16.

[153] E. g., for the Academy of Sciences efforts to restore foreign links, see Elizaveta D. Lebedkina, *Mezhunarodnyi Sovet nauchnykh soiuzov i Akademii nauk SSSR* (Moscow: Nauka, 1974), 35–57.

The 9th Congress of Soviets

THE 9TH CONGRESS OF SOVIETS, like its predecessor, did not discuss the GOELRO plan but heard a speech by Krzhizhanovskii, now head of the state planning organ Gosplan, and voted unanimously for thirteen resolutions on electrification in December 1921.[154] Compared with the December 1920 GOELRO plan, the milder 1921 resolutions reflected the new crises besieging the government—widespread famine in southeastern Russia, impending drought, continuing fuel shortages, and the withholding of international diplomatic recognition. These crises restricted economic growth, including the opening of the Kashira regional station, originally timed for the congress.[155]

The radical transformations and reconstructions of the future envisioned in the 1920 plan now stayed far in the background, like a wayward child at a family reunion. Downplaying the long-term maximum program of GOELRO for the immediate minimum program with its "enormous practical significance," Krzhizhanovskii declared that small-scale electrification offered the quickest approach to assisting the agricultural economy. Gosplan's Aleksandr A. Gorev, a leading electrical engineer, offered a vision of rapid rural electrification based on the conversion of the country's 45,000 water-powered mills into small electric stations, a far cry from the 112 regional stations and an echo of the alternatives raised and rejected at the 8th All-Russian Electrotechnical Congress.[156]

As at the electrotechnical congress, criticism centered not on the concept of planned electrification but on its distorted implementation and the cumbersome, overlapping organizations at all political levels. Although Krzhizhanovskii cited Western statements and the 8th All-Russian Electrotechnical Congress—a nonparty "impartial voice of the scientific-technical thought of the country"—as proof of the correctness of regional stations, he linked implementation more tightly than ever to domestic and foreign political and economic support.[157]

The final resolutions called for the construction of nineteen steam and eight hydroelectric regional stations, which would generate as much power as the original thirty first-priority stations.[158] The state

[154] *Deviatyi Vserossiiskii sezd sovetov: Stenograficheskii otchet (22–29 dekabria 1921)* (Moscow: VTsIK, 1922), 296–98; also "Postanovlenie SNK o plane elektrifikatsii Rossii," in Gladkov, ed., *K istorii plana*, 223–26.

[155] *Deviatyi Vserossiiskii Sezd*, 17.

[156] Ibid., 216, 218, 233–34.

[157] Ibid., 80, 215, 220–21.

[158] Ibid., 294–96.

would construct and operate the regional stations, but local development would determine the final size of each station. Depending on the economy, implementation would take ten to fifteen years. The congress deemed the construction of small and medium-size rural stations vital. Of special importance was the agricultural use of electricity in the southeast, hardest hit by famine. Railroad electrification fared less well. In contrast to the multitude of lines proposed by GOELRO, only three lines, linking the main industrial regions, were included.

In 1920, electrification became a state technology and a technological utopia in Russia. States have long promoted large technologies, such as railroads and canals, but the GOELRO plan was the most integrated, far-reaching, and propagandistic yet. Before GOELRO, the tsarist state supported the railroad for economic development; with GOELRO, the Soviet state harnessed electrification to transform the country socially, politically, and economically. For the electrical engineers, the harness was voluntary.

GOELRO represented a break with the old and a foundation for the new. Its utopian goals appeared utterly fantastic in the devastated Russia of 1920–21, yet the economic crises and the failure of the old state technology of railroads encouraged both bold radical thinking and its serious consideration by the country's new political leadership. Electrical engineers proposed two technological paths for post–civil war electrification: a centralized path of regional stations and a radical path of rapidly electrifying the countryside to link the peasantry with the city. A third, conservative path of expanding existing utilities found its strongest support among municipal governments, as Chapter 7 shows.

In the West as in Russia, World War I sparked great interest in large-scale engineering projects for social goals and contributed to the postwar popularity of technocratic thinking.[159] Russia was one of several countries where engineers and politicians tried to expand electrification along lines of bigness, nationalization, and efficiency. The three major unsuccessful Western electrification proposals of the 1920s were the unification of the Ruhr, "Giant Power" of Pennsylvania, and "Superpower" in the American northeast.[160] None of these

---

[159] Charles S. Maier, "Between Taylorism and Technocracy: European Ideologies and the Vision of Industrial Productivity in the 1920s," *Journal of Contemporary History* 5 (1970): 27–61; William E. Akin, *Technocracy and the American Dream: The Technocratic Movement, 1900–1941* (Berkeley: University of California Press, 1977), 4, 46.

[160] For the Ruhr, see Edmund N. Todd, "Technology and Interest Group Politics:

proposals was as wide-ranging geographically or economically as the GOELRO plan. Opposition from economic and political institutions, including railroads, utilities, and engineering societies, ultimately defeated these moderate proposals. More successful were the initial 1924 operations of the Walchenseewerk hydrostation and Bayernwerk transmission grid in Baveria, culminating nearly two decades of proposals, networking, and promotion by Oskar von Miller. Like Krzhizhanovskii, von Miller exemplified that rare engineer who could and would become deeply involved in politics."[161] Elsewhere, Great Britain nationalized its electric utilities in 1926 after intense political debate, and Ireland established the more modest Shannon scheme in the late 1920s.[162]

Like these Western proposals, the GOELRO plan was not so much a departure from contemporary trends as it was their directed extension. The theoretical currents that shaped GOELRO existed in the West, but in technology-fascinated Soviet Russia they became state policy. Why did Russia, instead of the more industrialized countries, adopt most completely the rationale of the machine age? How did Russia, which took so much from Western ideas, reach—on paper— technocratic goals first? The answer is revolution. The October revolution brought a regime to power that believed in the promise of machines to liberate.[163] The widespread economic devastation and depreciation of the old regime, combined with revolutionary hopes and expectations for the future, explain why this utopian plan of "electrofiction" became state policy in Soviet Russia but not in bourgeois Europe. Would a United States in a condition similar to that of Soviet Russia in 1920 have rejected Superpower packaged and politically supported like the GOELRO plan? One suspects not.

The West spawned and nurtured the basic ideas underlying GOELRO but neither the intense desire for nor the concomitant politi-

---

Electrification of the Ruhr, 1886–1930" (Ph.D. diss., University of Pennsylvania, 1984). For Giant Power, see Thomas P. Hughes, "Technology and Public Policy," *IEEE Proceedings* 64 (1976): 1361–71, and Bayla Singer, "Power Politics," *IEEE Technology and Society Magazine*, December 1988, 20–27. For Superpower, see Terry Kay Rockefeller, "The Failure of Planning for Electrical Power Supply: The Case of the Electrical Engineers and 'Superpower,' 1915–1924," in Joel A. Tarr, ed., *Retrospective Technology Assessment—1976* (San Francisco: San Francisco Press, 1977), 191–215.

[161] Hughes, *Networks of Power*, 334–50.

[162] Leslie Hannah, *Electricity before Nationalization* (Baltimore: Johns Hopkins University Press, 1979), 75–104; Michael J. Sheil, *The Quiet Revolution* (Dublin: O'Brien Press, 1984), 16–22.

[163] Remington, *Building Socialism*, 19, 117–18, 136–39; Stites, *Revolutionary Dreams*, 52, 145.

cal acceptance of technology to modernize.[164] GOELRO incorporated foreign ideas, but it integrated them into a uniquely Russian setting. World War I, the 1917 revolutions, and the civil war destroyed the existing political and economic order in Russia, unlike in Europe and North America, which changed under wartime pressures but did not collapse. The elimination by the Communist party of major sources of institutional opposition and the increase in central state control made electrification's rise to prominence possible.

In 1920, when military victory finally allowed a full debate about the country's future economic course, GOELRO seized the opportunity to fill a political vacuum. A small group of electrical engineers forged the necessary political, economic, and technological network of actors. Krzhizhanovskii was a technical entrepreneur who promised to provide the transition to a better future demanded by ideology. His status greatly contributed to GOELRO's approval, for he had high-level access and prestige in Communist party, government, and electrical engineering circles. Similar people existed in the United States, such as Superpower's William S. Murray and Giant Power's Morris Cooke, but their efforts, though painted on a regional canvas, were not crowned by success. One reason for Krzhizhanovskii's success was his ability to convince engineers and other specialists to work for the economic reconstruction of their country under a regime they disliked.[165] The majority of GOELRO planners were not Communists—Lenin described them as "almost all, without exception, opponents of Soviet power"—but intelligentsia who saw the new government removing the restraints of tsarism and supporting their modernizing mission.[166]

The GOELRO plan was not a technological fix; it was technological determinism writ large. It was boldness and brashness laced with hubris. The plan amalgamated many assumptions and goals, some contradictory, others openly based on aged and incomplete data. Ideological guidance came from a German, but Karl Ballod, not Karl Marx. The driver was the Communist party (Bolshevik), but the drivers were non-Bolsheviks operating by the principles espoused by a deceased "bourgeois" engineer, Grinevetskii. Most striking was GOELRO's heavy foreign reliance combined with revolutionary goals of transformation and more traditional goals of independence from

---

[164] Rainer Traub, "Lenin and Taylorism: The Fate of 'Scientific Management' in the (Early) Soviet Union," *Telos* 37 (Fall 1978): 87.

[165] Kulebakin, "Skromnyi vklad," *SRE*, 105.

[166] Lenin, *Collected Works*, 44: 51; Remington, *Building Socialism*, 124.

European domination. This reliance came from engineers who viewed their work as ultimus inter pares but were eager to catch up and from Bolsheviks impressed by Western technology and organization.

Electrification formed the plan's skeleton, but GOELRO's focus extended further to transform radically a backward country into a modern, Communist state by modern technology. This grandiose vision contained equally impressive subvisions, including eliminating the town–country gap, rationalizing labor, and creating new export industries. The essence of the GOELRO plan was not just command from above but centralized command based on a vision of the most rational exploitation of material and human resources with few insurmountable barriers. The NEP, however, made these assumptions problematic. Born in a centralized state, GOELRO would come of age in a decentralized economy. Obtaining an official imprimatur would prove trivial compared with implementation.

Krzhizhanovskii categorized GOELRO's critics into the right and left, conveniently placing GOELRO in the reasonable center. Both categories of criticism were correct: GOELRO remained in the tradition of Western development and Communist centralization. If the electrical engineers and the Communist party had fully supported the promise of small-scale electrification, that would have been the true revolution. Instead of building on mainstream Western lines, a uniquely Soviet decentralized model of electrification could have emerged. The social effects of electricity, especially in rural areas, might have been as revolutionary as GOELRO intended. In reality, the implementation of the GOELRO plan proved beyond the resources of 1920–21 and actually hindered the development of electrification through the mid-1920s. The road not taken demanded drivers other than the GOELRO engineers, whose professional training and visions concentrated on large centralized networks instead of smaller, isolated independent systems. Nonetheless, this plan of "electrofiction" represented a major step toward a permanent link between the state and specific technologies for broad-ranging goals. To look at Stalin's five-year plans and state promotions of large-scale technologies without looking at the GOELRO plan is to ignore the rich intellectual and technical soil in which they grew.

CHAPTER 7

The NEP Years,
1921–1926

ALTHOUGH MILITARILY SUCCESSFUL, war communism failed to transform Russia into a socialist society. By 1921, the worsening political
and economic situation demanded another approach. The Leningrad
strikes, Kronstadt sailors' rebellion, and Tambov peasant uprising
challenged the legitimacy of a government claiming to represent workers, soldiers, and peasants. The economic situation looked bleak: a
devastated transportation network, empty factories, rampant infla-
tion, recalcitrant peasants, famine and accompanying epidemics, high
unemployment, little trade with the West, and distinctly nonrevolu-
tionary, if not hostile, international relations.[1] In a reversal of war
communism, the government's NEP attempted to create an alliance
(*smychka*) between peasants and workers to restore the agricultural
economy and lead the peasants to socialism. The new policy legalized
small-scale capitalism and decentralized many industries while retain-
ing state control over heavy industry and foreign trade. State action
and market forces coexisted uneasily in a reviving mixed economy.

The shift of resources to more consumer- and peasant-oriented industries, reacquaintance with market forces, and severe financial constraints drastically changed the economic and political environment
for electrification.[2] A sharp political struggle over the future of electri-

---

[1] Paul Avrich, *Kronstadt, 1921* (New York: W. W. Norton, 1974); Stephen F. Cohen,
Bukharin and the Bolshevik Revolution: A Political Biography, 1888–1938 (New York: Vintage, 1973), 123–26.
[2] E. H. Carr, *The Bolshevik Revolution, 1917–1923*, vol. 2 (New York: Macmillan, 1952),
297, and *The Interregnum, 1923–1924* (London: Macmillan, 1954), 40.

fication dominated the 1920s. The foremost concern was control: who should direct electrification—the central government or local governments? The debate was not just about where to locate powerplants; it concerned the future nature of Soviet industrialization and society.

In principle, the NEP encouraged the decentralized development of local utilities: "From the viewpoint of electrification, it is vital to develop in every way possible private initiative and private enterprise in the area of electric construction."[3] Local governments considered that the NEP "opened new possibilities for the wide penetration into life by electricity through the construction of medium and small electric stations of local significance, *which were not foreseen* in the original [GOELRO] plan of electrification."[4] Urban utilities and their municipalities, the main losers under the GOELRO plan, used the NEP to regain and maintain their independence from Glavelektro, the successor to the ETO. Glavelektro countered that the state, with its "strongly planned and fully organized" approach, offered advantages of speed and cost over the municipal and private sectors. Furthermore, electrification's economic importance demanded state control, implemented by one center, rather than private monopolies.[5]

The centralization–decentralization debate increased in intensity as municipal governments and their state advocate, the NKVD, fought several battles over the control and direction of electrification with the proponents of centralization, Glavelektro and Gosplan, the state planning commission, and Elektrobank. By 1925, urban utilities had partially repudiated the GOELRO plan and called for a major redistribution of resources. Issues that appeared solely technical, such as inspection and standardization, played part of this sharp political struggle. The biggest debates were concerned with economics. Financing proved the utilities' most persistent problem and the central government's most potent weapon.

Electricity generation recovered faster than the rest of the economy, but the government's focus on regional stations impaired the recovery and expansion of utilities in the second and third tiers. The initial optimism for rural electrification weakened with a tide of problems encountered in transferring high technology into a low-technology environment. Some electrifiers refocused on regional stations; others

[3] V. L. Levi, "Rabota Glavelektro v usloviiakh novoi ekonomicheskoi politiki," *Voprosy elektrifikatsii*, 1922, nos. 1–2: 97.
[4] 'T', "Ob izmeneniiakh v plane elektrifikatsii, razrabotannom GOELRO," *Narodnoe khoziaistvo*, 1922, no. 1: 96, emphasis added.
[5] V. L. Levi, "Elektrosnabzhenie Rossii," TsGANKh f. 5208, op. 1, ed. kh. 69, 1, 4.

realized that, as with any technology, social and organizational changes had to accompany the successful diffusion of electric light and power into rural regions.

The actor network that created the GOELRO plan lost some of its effectiveness as promotions increased the responsibilities of its leaders, while Lenin's death and instability of the Glavelektro chair reduced its power. The first half-decade of the NEP demonstrated the state's political reinforcement of one line of technological development, regional stations, at a time when conventional economics and urban governments urged a redeployment of resources elsewhere. Ironically, Glavelektro beat off these challenges but lost its priority to the demand of the state and party for accelerated industrialization.

Organization of State Electrification Agencies

THE PULL of central political support and the push of increasing demand shaped the bureaucratic evolution of electrification. The early 1920s saw the growth and maturation of new organizations and responsibilities at the national, regional, and local levels. Although electrifiers did not amass great administrative strength, electrification nonetheless became solidly embedded in the Soviet state apparatus. Most important were the failure to create a people's commissariat of energy, the growing power of central planning and control, the rise of central financial organs, and the creation of local companies.

In 1920, the state organizational structure remained fairly simple. The VSNKh housed the KGS and ETO/Glavelektro. The KGS's Elektrostroi handled construction and, through the TsES, technical advice. GOELRO assumed that it would represent electrification in a future state planning commission and that the ETO and Elektrostroi would implement the plan. GOELRO justified its continued existence on the grounds that the task of electrification remained unfinished. Prewar utilities would need replacement and the smaller stations should be integrated into a national grid.[6] Electrification still required much planning, and GOELRO intended to do it.

GOELRO's assumptions proved unrealistic. By 1924, a series of battles over the course of electrification had produced a more complex organizational matrix that diffused responsibility and increased duplication, the worst of both worlds. The 1921–22 battle over the future of

[6] 14 December and 3 November, *Trudy GOELRO: Dokumenty i materialy* (Moscow: Izdatelstvo sotsialno-ekonomicheskoi literatury, 1960), 192–94, 183–84.

the ETO demonstrated the subordination of electrification to more powerful state interests. On 21 December 1921, the VSNKh ordered the newly formed Glavelektro, the ETO successor, to submit organizational proposals to fulfill the GOELRO plan.[7] Four months later, Glavelektro submitted three options: a commissariat of energy, a commissariat of electrotechnology, and a continuation of Glavelektro. The major differences were the degrees and centralization of authority, vertical integration, and independence from the VSNKh.

Glavelektro favored a commissariat of energy to control the entire energy cycle from coal mine to consumption, unifying Glavelektro, Elektrostroi, and the Main Fuel Administration into a center above the VSNKh and answerable only to the powerful SNK. Controlling all energy would provide "the state exceptional advantages for administering all the economy," since the government could "prevent all those activities that can lead to violations of the general state economic plan." This proposal implied economic management by control of energy and a return to a command economy. More important, the proposed commissariat would create a new political center for industrialization and threaten the authority of the VSNKh, the Commissariat of Transportation, and other state organs.[8] Such ambitious schemes proved politically impossible in 1922. The rejection of a commissariat of energy or milder commissariat of electrotechnology in favor of the existing Glavelektro indicated the limits of the political power of the electrical engineers and the strength of the established state structure.

Officially the VSNKh Main Electrotechnical Administration, Glavelektro handled the electrotechnical industry, electricity supply, and, with the transfer of the TsES from Elektrostroi, grand advice.[9] Glavelektro directly controlled the three largest centers of electrical consumption, Moscow, Petrograd, and Baku. Although it eventually lost direct control over smaller stations, it maintained its influence by establishing standards for equipment, tariffs, and operations, and, most important, by coordinating utility financing with Gosplan and Elektrobank.

Five men headed ETO/Glavelektro from 1920–26: Gleb M. Krzhizhanovskii, Valerian V. Kuibyshev, Abram Z. Goltsman, Leon Trotsky, and Isaak E. Korostashevskii. Of the five, only Krzhizhanov-

---

[7] TsGANKh f. 3700, op. 1, ed. kh. 6, 1, 62–63.

[8] Ibid., 10–14.

[9] V. V. Kuibyshev, "Sostoianie elektrotekhnicheskoi promyshlennosti i elektrosnabzheniia R.S.F.S.R. k nachalu 1922," *Voposy elektrifikatsii*, 1922, nos. 1–2: 49.

skii was a prominent electrical engineer. Kuibyshev and Trotsky were professional party members who served only short stints and whose interests lay with the party. Korostashevskii served as Glavelektro vice-chairman before and after his period as director. Goltsman was a Bolshevik worker with a polytechnic education who served on the central committee of the metallurgy union. Irrespective of the individuals, five heads in six years provided neither firm leadership, continuity, nor skill in bureaucratic politics.[10] The political appointees did not stay long enough to influence the VSNKh, and the electrical engineers played a subordinate role to their political superiors.

Possibly the most unexpected director was Trotsky, appointed in May 1925 following his January dismissal as commissar of war. Together with his appointment to head the concessions commission and the VSNKh science-technology section, this selection marked a major stage in his gradual political defeat by Stalin. Although placing him under Felix Dzerzhinskii, VSNKh chairman and founder of the Cheka, the Soviet secret police, these appointments allowed Trotsky to pursue his interests in industrial planning, including scientific management.[11] At the June 1925 opening of the Moscow Thermo-Technical Institute named for the heat committee's Kirsh and Grinevetskii, Trotsky called for "Ramzinists" to explore and promote rationalized fuel use and industrial organization because "scientific technology is one of our most important weapons of our state self-assertion in the world struggle."[12] This connection with Trotsky may have contributed to Ramzin's involuntary participation in the 1930 Industrial party trial.[13] In 1925, Trotsky adopted a comparative approach in industrial planning by organizing a body of U. S. experts, augmented by a similar body of Germans, to guard the Dniepr hydrostation planning from defective estimates and to attract foreign involvement.[14]

[10] For a discussion of bureaucratic politics in another context, see Harvey M. Sapolsky, *The Polaris System Development: Bureaucratic and Programmatic Success in Government* (Cambridge: Harvard University Press, 1972), 242–44.

[11] Rainer Traub, "Lenin and Taylorism: The Fate of 'Scientific Management' in the (Early) Soviet Union," *Telos* 37 (Fall 1978): 89; Ronald Segal, *Leon Trotsky: A Biography* (New York: Pantheon, 1979), 290–91; Leon Trotsky, *My Life* (New York: Charles Scribner's Sons, 1930), 518.

[12] L. Trotsky, "Nauchno-tekhnicheskaia mysl i sotsialisticheskoe khoziaistvo," *Izvestiia*, 2 June 1925, 3.

[13] Kendall E. Bailes, *Technology and Society under Lenin and Stalin: Origins of the Soviet Technical Intelligentsia, 1917–1941* (Princeton: Princeton University Press, 1978), 96–121.

[14] Trotsky, *My Life*, 519; "Soviet Experts in the United States," *Russian Review*, June 1926, 148.

Prior to 1925, Trotsky had viewed regional electrification as a long-term goal with colossal significance for integrating the peasants into a single economic plan. His interest predated the GOELRO plan, but, except for his support of Shatunovskii's 1921 unrealistic hydroelectric proposal for Petrograd, Trotsky did not favor an aggressive approach to electrification. Although important, electrification was a ten-year plan, and other industries—particularly transportation—demanded more immediate priority.[15]

Trotsky claimed that he worked energetically at Glavelektro, but his political baggage disrupted his new duties: "Every practical step that I took gave rise to a complicated intrigue behind the scenes." Equally important, "it became practically impossible for the institutions under my direction to obtain the necessary wherewithal. People working there began to fear for their futures, or at least for their careers."[16] According to Valentinov-Volsky, a high-level VSNKh official, Trotsky lost the great initial interest of the VSNKh staff by his lack of patience, attention to too many issues, and leisurely work habits.[17] Instead of vigorous leadership, Glavelektro suffered the double handicap of an unmotivated director in political disgrace. In January 1926, Trotsky asked to be relieved of his VSNKh assignments. His vice-chairman, Korostashevskii, replaced him.[18]

Electrification was affected not only by changes in Glavelektro chairmen but from an expansion of bureaucracy. Elektrostroi remained the state construction organ under the KGS, now renamed the Main Administration for State Construction. At the state level, the NKVD Main Administration for Cities (GUKKh, Glavnoe Upravlenie Kommunalnykh Khoziaistv) represented urban interests. Like the tsarist MVD, the GUKKh's planning commission reviewed city projects and worked with Gosplan.[19]

Above Glavelektro and the GUKKh stood Gosplan. Instead of becoming part of Gosplan, established in February 1921 to develop a unified plan for the entire economy, GOELRO dissolved into the new

---

[15] Leon Trotsky, *Sochineniia* (Moscow-Leningrad: Gosizdat, 1927), 15: 134, 234–36; G. M. Krzhizhanovskii, "Perspektivy elektrifikatsii," *Planovoe khoziaistvo*, 1925, no. 2: 6; Jan M. Meijer, ed., *The Trotsky Papers* (The Hague: Mouton, 1971), 2: 446–51; Trotsky, *Osnovye voprosy promyshlennosti* (Moscow: Ekonomicheskaia Zhizn, 1923), 33–34.

[16] Trotsky, *My Life*, 520.

[17] N. Valentinov (Volsky), *The New Economic Policy and the Party Crisis after the Death of Lenin* (Stanford: Hoover Institution Press, 1971), 199–200, 219–23.

[18] "Prikaz 312," *Sbornik postanovlenii i prikazov po promyshlennosti*, 1926, no. 8: 60.

[19] "Khronika Tsentra," *Kommunalnoe delo*, 1922, no. 2: 44–45; 1924, no. 5: 32–33; "Na mestakh," *Kommunalnoe delo*, 1924, no. 1: 73.

organization.[20] Gosplan relied heavily on GOELRO members, including Krzhizhanovskii, its first president; Osadchii, a vice-president; Aleksandr A. Gorev, a presidium member; and at least eleven other people.[21] Within Gosplan, the section on energy handled electrification.[22] The GOELRO veterans had greater authority and power than before, but their responsibilities and priorities changed also. The initial political emphasis on the short term and responsibilities far beyond electrification meant that implementation of the GOELRO plan would not receive top priority despite the high posts of its founders. The electrifiers became captives of their positions.

The greatest growth occurred in planning organs as each agency established its own cadre of planners to work with Gosplan. Both Glavelektro and the TsES established planning sections in 1921–22. At the urging of the Gosplan section on energy, the VSNKh created in July 1922 a planning commission for electrification, Elektroplan, which united the planning activities of Glavelektro, Elektrostroi, and the TsES; in early 1923, the activities of the GUKKh were added, belatedly reflecting the transfer of local utilities from Glavelektro to the GUKKh. The TsES abolished its planning section in mid-1924 on grounds that its representatives in Elektroplan rendered the section redundant, but it then created a new section for regional stations.[23]

With four central planning organs and a state-approved plan, electrification suffered not from a deficit of thought but a lack of meaningful action. These sections repeatedly sent plans and proposals back and forth, seeking unanimous approval and diverting disputed issues elsewhere.[24] Elektroplan was the most active organ, followed by the Gosplan section on energy. These planning bodies did not contribute to the realization of electrification, only to its bureaucratization, yet another legacy of the tsarist government.

A major focus of early planning consisted of devising operating and

[20] V. G. Smoliakov, *Voprosy gosudarstvennogo stroitelstva v resheniiakh IX-ogo Vserossiiskogo sezda sovetov* (Moscow: Izdatelstvo Moskovskogo universiteta, 1962), 38–44.

[21] Including Ivan G. Aleksandrov, Kogan, Krug, and Shulgin; *Trudy GOELRO*, 259–74.

[22] "Iz zhizni," *Elektrichestvo*, 1922, no. 2: 54–55. An early task was sending local governments copies of the GOELRO plan.

[23] P. M. Tikodeev, "Komissiia po osvetitelnoi tekhnike pri TsES," *Elektrichestvo*, 1923, no. 4: 183–84; "Iz zhizni," *Elektrichestvo*, 1922, no. 2: 56–57; P. S. Osadchii, "Tsentralnyi elektrotekhnicheskii sovet," *Elektrichestvo*, 1922, no. 3: 2–5; "Khronika," *Elektrichestvo*, 1923, no. 4: 236; 1924, no. 8: 414–15; P. S. Osadchii, ed., *Materialy k XIX-oi sessii plenuma TsESa v marte 1925 g.* (Moscow: Glavelektro, 1925), 3.

[24] E. g., the siting of the Shatura substation; "Khronika," *Elektrichestvo*, 1924, no. 5: 289.

construction plans for Petrograd and Moscow on one-year and five-year bases; these constantly underestimated demand and overestimated supply.[25] Elektroplan and Gosplan actually exerted decisive influence in settling the constant arguments within the Moscow and Petrograd utilities over the appropriate mix of stations for generation.[26] The planners responded as well as initiated, using their commanding central position to coordinate separate activities.[27] One benefit of the emphasis on planning was the creation of far more comprehensive statistics. As lines of authority firmed and the outlook for regional stations improved, the central planning organs devoted more resources to areas outside the first tier.

In contrast with the growth of the state electrification bureaucracy, the VI Section and the military did not regain their prerevolutionary roles. Although the VI Section of the Russian Technical Society, shorn of its "Imperial" designation, and the Moscow Society of Electrotechnicians resumed activities in 1921–22, the professional electrical engineering associations never regained their prerevolutionary prominence or wartime initiative.[28] Instead, the TsES and planning organs now advised and guided cities, while society members proposed ideas within rather than to the government. In achieving state-sponsored electrification, the VI Section lost its independence. Before 1917, it was a junior partner and occasional adversary of the government; now it functioned increasingly as a branch of the government.

During World War I, the military again played an important role because of its industrial needs and authority to allocate resources. It could protect skilled workers from mobilization, obtain priority in materials and transport for factory orders, and transfer surplus military equipment to generate electricity elsewhere. This equipment proved marginally important during the civil war, but as better equipment replaced the old equipment the military retreated from an active role in electrification.

The first electrotechnical journals published since 1918 appeared in 1922, a sign of economic recovery. *Voprosy elektrifikatsii* (Questions of

[25] "Iz zhizni," *Elektrichestvo*, 1923, no. 1: 55; "Biulleten," *Elektrichestvo*, 1928, nos. 1–2: 35; 1930, no. 1: 50–51, no. 3: 159.

[26] "Khronika," *Elektrichestvo*, 1923, no. 4: 236, no. 10: 530; A. A. Gorev, "Elektrifikatsiia SSSR," *Planovoe khoziaistvo*, 1925, no. 2: 169.

[27] E. g., the Urals planning organ combined proposed stations for a coal mine and a factory into one larger station at the coal mine; "Kommunalnye predpriiatiia," *Kommunalnoe delo*, 1925, no. 8: 59–60.

[28] "Iz zhizni," *Elektrichestvo*, 1922, no. 1: 52; 1923, no. 3: 165.

Electrification) appeared only once before replaced by *Elektrichestvo*, now the organ of Glavelektro as well as the electrical engineering community. As before, the journal served as a major conduit for information from the West and the government. Significantly, coverage of urban electrification shifted to *Kommunalnoe delo* (Communal Affairs), the NKVD journal for municipal governments. Another new journal, *Elektrifikatsiia*, provided information for rural electrification.

The state agencies promoting electrification under the NEP were not what Krzhizhanovskii and other electrifiers desired. Instead of commanding positions, they were subordinated to other state authorities; instead of clear-cut lines of authority, bureaucratic confusion resulted. Confusion grew not just from administrative flux but from very real uncertainty among government officials as well as engineers about what to do and how. Nonetheless, within the realm of electrification Glavelektro successfully forged an alliance with Gosplan and Elektrobank to defeat alternative programs. Against the interests of other parts of the state, particularly heavy industry, and the Communist party, Glavelektro faired less well in the larger debates about economic development.

The Implementation of the GOELRO Plan

THE IMPLEMENTATION of the GOELRO plan differed greatly from its creation. Instead of generating a well-funded, well-organized, and centralized program to build regional stations, the drive to electrify split into competing directions, which vied for resources. The NEP shift of the political and economic environment toward decentralization and short-term, profitable operations was the major causative factor. Inadequate funding chronically hindered electrification, especially in the slow economic recovery of 1921–22.

Faced with continuing "bureaucratic irresponsibility and muddles" and shortages of equipment, materials, and food, the government in June 1921 stopped construction on all regional stations except for projects promising short-term results, like the temporary 5-MW Shatura station. Other projects, including railroad electrification, experienced similar delays.[29] These cutbacks grew from an overextension of

[29] Vadim A. Smolianinov, "Velikii stroitel," *Sdelaem Rossiiu elektricheskoi* (Moscow: Gosenergoizdat, 1961), 57–75; "Gosplan za polgoda svoei raboty," *Narodnoe khoziaistvo,* 1921, nos. 11–12: 47; 1 June 1921 Council for Labor and Defense resolution, Iuri A. Gladkov, ed., *Razvitie elektrifikatsii sovetskoi strany, 1921–1925 gg.: Sbornik dokumentov i*

resources caused by the simultaneous construction of too many stations. Since total investment did not increase sufficiently, funding was greatly inadequate to complete all the stations. Limited resources were spread too thin, the result of bad planning. At a time when reconstruction demanded optimal use of financial, material, and human resources, this waste delayed recovery. Partially constructed stations contributed nothing to economic growth; indeed, they postponed growth by absorbing resources better used elsewhere.

As obstacles mounted, expectations fell. In February 1922, a pessimistic Krzhizhanovskii forecast that the construction of the twenty-seven regional stations would demand ten to twenty years, a potential doubling of the original goal. In May, a major article claimed ongoing construction on thirteen of the twenty-seven stations. Half a year later, only ten stations were so described, work on two had stopped, and two other stations remained in the planning stage.[30] According to V. V. Kuibyshev, Krzhizhanovskii's successor at Glavelektro, large-scale construction presented a "not especially delightful picture."[31]

Divided authority, poor organization, and local–center conflicts hindered the allocation of inadequate resources. In 1921, Elektrostroi controlled construction of regional stations with the significant exception of stations already started. The VSNKh presidium directly controlled Volkhov construction, and Glavelektro managed Shatura construction. Not until 1924 did Glavelektro receive responsibility for construction of all regional stations, excluding the Volkhov.[32]

One problem generic to the construction of regional stations and other large-scale projects was the poor living conditions. Overcrowding, illnesses, and infestations of cockroaches and bedbugs so bad that "night without light is unsafe from them" were not unusual.[33] The terribly inadequate living conditions at Kizel, documented by an inspection team sent from the Commissariat of Labor in 1922 because of a typhus epidemic, hurt the health of the labor force.[34] As the large-

---

materialov (Moscow: Gospolitizdat, 1956), 43–44; 'T', "Ob izmeneniiakh v plane elektrifikatsii," 95–100.

[30] "Inostrannyi kapital i elektrifikatsiia Rossii," *Trud*, 15 February 1922, 4; A. Gorev, "Raboty po osushchestvleniiu elektrifikatsii v Rossii," *Voprosy elektrifikatsii*, 1922, nos. 1–2: 21; K. V. Bulgakov, "Kratkie svedeniia o glavneishikh elektrostroitelstvakh v Rossii," *Elektrichestvo*, 1922, no. 3: 17.

[31] Kuibyshev, "Sostoianie elektrotekhnicheskoi," 43.

[32] "Ekonomicheskii otdel," *Pravda*, 31 January 1922, 4; *Obzor sostoianiia rabot po krupnomu elektrostroitelstvu na 1ogo oktiabria 1925 g.* (Moscow: Glavelektro, 1926), 18.

[33] E. g., Kashira and Volkhov; "Vnimaniiu Glavelektro," *Pravda*, 19 July 1924, 5.

[34] M. Israel, "Eshche o Kizelstroe," *Uralskii rabochii*, 16 July 1923, 6.

Kashira regional station. Courtesy of the Soviet Polytechnic Museum.

scale projects of the Stalin era and the tsarist trans-Siberian railroad demonstrated, poor living and working conditions were the rule, although conditions in the tsarist era were better.[35] One contributing factor was the siting of regional stations in undeveloped areas, which necessitated first the establishment of construction industries, including sawmills and brickworks, to build the town needed to construct the plant.[36]

The Ural Kizel plant exemplified the problems of translating GOELRO's vision into reality. The Kizel project suffered from a scan-

[35] E. g., John Scott, *Behind the Urals: An American Worker in Russia's City of Steel* (Boston: Houghton Mifflin, 1942); Anne D. Rassweiler, *The Generation of Power: The History of Dnieprstroi* (New York: Oxford University Press, 1988); Steven G. Marks, *The Road to Power: The Trans-Siberian Railroad and the Colonization of Asian Russia* (Ithaca: Cornell University Press, 1991), chap. 9.

[36] For Shatura, see Allan Monkhouse, *Moscow, 1911–1933* (London: Victor Gollancz, 1933), 117.

dal of equipment shipped in 1919 from the Oranienbaum electric station but still absent from Kizel in 1921. More mundane but no less serious were the death of the chief engineer, desperate living conditions, an epidemic, inadequate work clothing, poor communications, too few horses, and lack of assistance from Elektrostroi.[37] The Kizel station's continued difficulties threatened the area's mining industry. In a now-familiar tale, shortages of money in 1922–23 also delayed construction. The station finally opened with 6 MW (15 percent of its planned capacity) in spring 1924, powered by the two 3-MW generators from Oranienbaum.[38]

The overall situation improved in 1923 as a growing economy, the end of the famine, increased state funding, and greater access to materials, equipment, and fuel eased many shortages. Repairs and reconstruction of existing stations resulted in greater availability of equipment, and three years of experience had produced a more mature and seasoned set of administrators. Nonetheless, projects moved slowly toward completion, with initial capacity significantly below planned capacity (see Table 7.1).

The value of regional stations began to improve drastically in 1926, when three regional stations— Volkhov, Nizhegorod, and Shterov— and the main Shatura station began operations, consummating the work of several years. The 164 MW of new construction, although only a tenth of the capacity planned by GOELRO, nonetheless equaled the prewar capacity of all second- and third-tier stations. Of the seven regional stations generating electricity in December 1926, one (Red October) had prewar origins and served Leningrad like the just opened Volkhov, and two (Kashira and Shatura) served the capital, leaving only the Kizel, which was more an industrial station than a regional station, Shterov, and Nizhegorod stations providing power outside the Moscow–Leningrad industrial nexus.[39] Those three stations did not generate even 5 percent of the electricity of first-tier and regional stations in 1926–27.[40]

Compared with the goals and promises of GOELRO, rhetoric outpaced results. GOELRO's minimum plan produced incremental, not

---

[37] "Kizelstroi," *Uralskii rabotnik*, 12 January 1923, 4–5; V. Avanesov, "Elektrifikatsiia na Urale," *Ekonomicheskaia zhizn*, 15 January 1922, 1; Vasilii I. Steklov, *Lenin i elektrifikatsiia* (Moscow: Nauka, 1975), 140.

[38] "Kizelstroi," *Uralskii rabotnik*, 12 January 1923, 5; *Obzor sostoianiia rabot*, 35; "Rezoliutsii Pervoi Vsesoiuznoi konferentsii po elektrosnabzheniiu," *Elektrichestvo*, 1924, no. 9 (conference supplement): 4.

[39] A. Gorev, "Planovaia elektrifikatsiia," *Pravda*, 4 January 1925, 1.

[40] "Biulleten," *Elektrichestvo*, 1930, no. 1: 50–51.

Table 7.1. Capacity of regional stations, 1924–26

Station	GOELRO planned (MW)	1924 actual (MW)	1925 actual (MW)	1926 actual (MW)
Kashira	60	8	12	12
Shatura	40	5	5	32
Volkhov	54	—	—	54[a]
Red October[b]	60	6	10	20
Kizel[c]	40	—	6	6
Nizhegorod[d]	40	—	—	20
Shterov	60	—	—	20
total[e]	314 (68)	19 (100)	33 (82)	164 (72)

Sources: GOELRO and 1924, A. A. Gorev, "O sostoianii rabot po planovoi elektrifikatsii SSSR," *Elektrichestvo,* 1924, no. 12: 578, 1925, Glavelektro, *Obzor sostoianiia rabot po krupnomu elektrostroitelstvu na 1ogo oktiabria 1925 g.* (Moscow, 1926), 6, 1926, "Biulleten," *Elektrichestvo,* 1927, no. 1: 43.
[a] Volkhov did not open until December 1926.
[b] Formerly Utkina Zavod.
[c] Renamed Gubakha.
[d] Renamed Balakhnin.
[e] Percentage of Moscow and Leningrad plants to all stations in parentheses.

revolutionary, improvements. Engineers measured their progress by electrifying a large factory in Nizhni-Novgorod, extending transmission lines to towns near the Kashira station, and understanding mining needs to design the Kizel station better. In an inexpensive approach to spreading electric light and power, low-voltage transmission lines electrified areas outside seventy-seven factories.[41] The economy's recovery on traditional lines, however, derailed efforts to create new industrial regions. In the Northern Region, predicted demand rose faster than predicted supply, whereas Siberia grew too slowly to justify a regional station.[42]

Regional stations remained the government's priority. State funding skewed heavily toward them—of the 258 million rubles directly invested in utilities from 1920 to 1926, 229 million (89 percent) went to regional stations and only 29 million to urban and rural stations. Of the 229 million rubles, the capital-intensive Volkhov project consumed 93 million (41 percent), though its 54 MW provided only one-

[41] "Biulleten," *Elektrifikatsiia,* 1928, no. 11: 39.
[42] "Rezoliutsii Pervoi," 3–5.

third of the 164 MW installed in 1926.[43] The "white coal" revolution cost 2.5 times more per megawatt than other fuels (1.7 to 0.7 million rubles/MW).[44]

In 1926, Krzhizhanovskii called the delay in hydrostation construction electrification's major problem.[45] The record was poor: the Volkhov station finally opened in late 1926, eight years after the ETO started construction, three years after the original 1923 goal, and two years from GOELRO's 1924–25 goal.[46] The Volkhov was a symbol of the electrified future as much as Leningrad's main power source. Nonetheless, July 1925 found construction finished only in the rough. Swiss engineers began installing the turbines only in August 1925. Workers completed the transmission line to Leningrad in November 1926 and the hydrostation officially opened on 19 December 1926.[47]

The only other hydrostation under construction was the 11 million ruble Zemo-Avchalsk in Georgia, which grew from local initiative with some state financial assistance. Planning of the 13-MW station started in June 1922, construction commenced in early 1923, and operations began in June 1927, two years later than scheduled.[48] Costs had also significantly exceeded the original estimates. Mikhail D. Kamenskii, secretary for the short-lived Union of Electrotechnicians and ETO vice-chairman, had warned in 1920 that, based on Western experience, the construction of a hydrostation demanded six to ten years and cost three to five times more than thermal stations.[49] Unfor-

---

[43] Elektrobank, *Finansirovanie elektrokhoziaistva (dva goda raboty Elektrobanka)* (Moscow-Leningrad: Promizdat, 1927), 10–11. The sum excludes civil war expenditures of 15–20 million rubles; "Elektrosnabzhenie i elektrostroitelstvo," *Ekonomicheskaia zhizn*, 26 February 1925, 5.

[44] "Biulleten," *Elektrichestvo*, 1927, no. 1: 43; A. Barilovich, "Plan elektrifikatsii," *Elektrifikatsiia*, 1925, nos. 11–12: 4.

[45] "Perspektivy elektrifikatsii SSSR," *Ekonomicheskaia zhizn*, 12 December 1926, 1.

[46] *Plan elektrifikatsii GOELRO* (Moscow: Politizdat, 1955), 277; "Volkhov," *Izvestiia Elektrotresta*, 1920, no. 3: 5.

[47] "Po soiuz respublik," *Izvestiia*, 30 July 1925, 4; "Zavtra Volkhovskaia stantsiia dast tok k zavodu," *Leningradskaia pravda*, 19 November 1926, 4; "K otkrytiiu Volkhovskoi gidroelektricheskoi stantsii im. Lenina," *Krasnaia gazeta*, 19 December 1926, 2; Steklov, *Lenin i elektrifikatsiia*, 131–36.

[48] "V presidiume Gosplana," *Ekonomicheskaia zhizn*, 5 May 1925, 5; A. G. Kolossov, S. A. Kukel, and I. A. Skavani, "Obzor elektrosnabzheniia S.S.S.R. k kontsu 1923 goda," *Materialy po elektrosnabzheniiu S.S.S.R.* (Moscow: VSNKh, 1924), 19; "Khronika kommunalnoi tekhniki," *Kommunalnoe delo*, 1925, nos. 11–12: 85–86; N. A. Khachaturov, "Zemo-Avchalskaia gidroelektricheskaia stantsiia," *Elektrifikatsiia*, 1926, no. 8: 27–31; Steklov, *Lenin i elektrifikatsiia*, 145–48.

[49] M. D. Kamenskii, "Svirskaia i Volkhovskaia gidro-elektricheskie stantsii i realnye vozmozhnosti obespecheniia petrogradskogo raiona na blizhaishchee vremia," Ts-GANKh f. 5208, op. 1, ed. kh. 69, 95.

Lenin and electrification. Volkhovstroi produces power! Communism is Soviet power plus electrification. Courtesy of the Hoover Institution.

tunately for the Soviet economy, his prediction proved more accurate than GOELRO's.

The heavy investment in the Volkhov raises the question of whether the capital-short country should have focused its resources elsewhere. Alternatives for Leningrad were, however, few. In 1920, Kamenskii had proposed expanding existing thermal stations and using British coal for most efficient operations.[50] But returning to pre-war dependence on British coal was politically impossible domestically, so the easiest technological choice was eliminated. Shipping southern oil and coal would strain the transportation system and contradict the concept of regional autarky. Local low-quality peat did supply Petrograd, but never at the levels necessary. Furthermore, the cult of "white coal" was deeply entrenched in the mythos of electrification. Against the vision of massive dams feeding the country's vibrant industries tomorrow, what were high capital costs and shortages today?

The answer also depended on who was paying. Gosplan's Gorev decried the opposition to the long-term investment necessary to create a centralized energy supply.[51] But industries could not afford to wait for the future. Delays of regional stations forced electricity-dependent industries to upgrade and expand their own stations. Delays and inadequate capacity in the "long-constructed and long-awaited" Kizel station forced the city and the Karabash copper factory to add 2.5 MW to their stations.[52] These expansions, often done at the last minute, were costly and resulted in equipment purchases that soon became surplus. They also inspired the desire to minimize future dependence on outside electricity.

Promises of regional stations remained mostly that during the early NEP years. The major exception was Moscow; Leningrad proved only a partial exception. In a continuation of prerevolutionary activity, the first half-decade of the GOELRO plan had increased the electricity supply for the first tier, not for the country as a whole.

The First Tier

GOELRO's FIRST YEARS WERE, to a great extent, the history of the electric networks of Petrograd/Leningrad and Moscow, which pro-

---

[50] Ibid., 95.

[51] A. A. Gorev, "Elektrifikatsiia SSSR," *Planovoe khoziaistvo*, 1925, no. 2: 177.

[52] N. Birin, "Elektrifikatsionnye neuviazki," *Ekonomicheskaia zhizn*, 3 January 1926, 5.

duced nearly half the country's utility-generated power and consumed the majority of resources and investment. The government's concentration of resources on the first tier enabled these utilities to recover and expand more rapidly than those of other cities. Moscow and Baku surpassed their 1916 highs in 1922–23, four years before Leningrad, because their utilities directed investment into traditional technologies and locally fueled regional stations (see Graph 7.1).[53] Moscow continued to expand rapidly with new regional stations while Baku, which did not add any new stations, grew more modestly. Leningrad suffered longer because the bulk of its investment went into the Volkhov hydrostation and because the Soviet government chose not to resume imports of British coal.

The unheralded success of postrevolutionary electrification was Baku, which generated more electricity than Leningrad until 1927. Baku's two central stations had suffered catastrophically from worn-out equipment, the lack of replacement equipment, shortages of trained personnel, and overloading.[54] Aided by new insulators and other imported equipment, the two-station network reached prewar levels in 1919, in sharp contrast to Moscow and Petrograd, and rapidly expanded in the mid-1920s. Under the rubric of rationalization, the Azerbaidzhian oil trust, Azneft, increased the share of electric motors in oil fields from 54 percent in 1920 to 83 percent in 1926.[55] This major accomplishment of the Soviet government received little attention, in large part because regional stations were not involved.

Nowhere was the gap between the promise and the reality of GOELRO greater than in Petrograd. Its utilities suffered more than Moscow's because of the variability of the low-quality fuel burned and shortages of spare parts and equipment: in 1922, only half of Petrotok's installed capacity functioned. The 1st State Electric Station (formerly the 1886 Company station) had only 22 MW of working turbines, compared with a theoretical 45 MW, and seven of the nine working turbines needed blade changes. Similar problems afflicted

[53] "Biulleten," *Elektrichestvo*, 1928, nos. 1–2: 35; 1930, no. 1: 50–51, no. 3: 159; A. A. Kotomin, "Deiatelnost Leningradskikh obedinennykh gosudarstvennykh elektrostantsii (Elektrotok)," *Elektrichestvo*, 1925, no. 5: 329.

[54] Azneft, *Bakinskaia neftianaia promyshlennost za tri goda natsionalizatsii (28 maia 1920–28 maia 1923)* (Baku: Azneft, 1923), 21–26; Azneft, *Azerbaidzhanskaia neftianaia promyshlennost za 10 let natsionalizatsii, 1920–1930* (Baku: Azneft, 1930), 61.

[55] "Elektrifikatsiia Bakinskogo raiona," *Elektrichestvo*, 1923, nos. 5–6: 306; William A. Otis, "The Petroleum Industry of Russia," Trade Information Bulletin No. 263 (Washington, D.C.: Dept. of Commerce, 1924), 20; Azneft, *Azerbaidzhanskaia neftianaia promyshlennost*, 111.

Graph 7.1. First-tier electricity generation, 1913–27

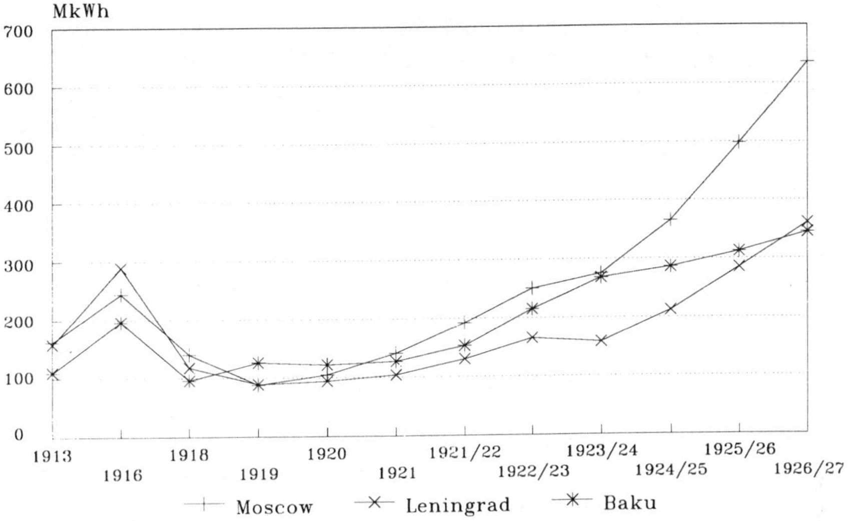

MkWh

Sources: "Biulleten," *Elektrichestvo*, 1923, no. 9: 460, 465; 1928, nos. 1–2: 35; 1930, no. 1: 50–51, no. 3: 159; A. A. Kotomin, "Deiatelnost Leningradskikh obedinennykh gosudarstvennykh elektrostantsii (Elektrotok)," *Elektrichestvo*, 1925, no. 5: 329; S. P. Stafrin, "Rabota elektricheskikh stantsii Moskovskogo raiona za 1922 g.," *Elektrichestvo*, 1922, no. 1: 36; Gudrat Ia. Abdulsalimzade, *Osushchestvlenie Leninskogo plana elektrifikatsii v Azerbaidzhane* (Baku, 1968), 36, 81.

boilers and auxiliary equipment.[56] Major repairs and new boilers for the former 1886 Company station in 1924–25 restored its capacity to prewar levels, but inadequate boiler capacity constrained output through the 1920s.[57] The incompatibility of voltage and frequency of the four utilities hindered citywide operations until the completion of the 1915 unification plan, originally scheduled for 1917–19, in 1925.[58]

Organizationally, the ETO operated the 1st Station and the city's

[56] A. V. Vulf, "Elektrosnabzhenie Petrograda," *Elektrichestvo*, 1922, no. 1: 4–13. For a turbine-by-turbine survey, see A. A. Kotomin, "K voprosu o perspektivakh elektrosnabzheniia Petrograda i ego okrestnosti v sezone 1921–22 g.," TsGANKh f. 5208, op. 1, ed. kh. 69, 123–29, and "O remonte turbin, proizvedennom na petrogradskikh elektrostantsiiakh," *Elektrichestvo*, 1923, nos. 7–8: 358–67.

[57] In October 1927, only 96 MW of boiler capacity functioned compared with 129 MW of available turbine capacity; Elektrotok, *Statisticheskii spravochnik 'Elektrotoka' 1913–1928* (Leningrad: Elektrotok, 1929), 10–11.

[58] I. A. Skavani, "Elektrosnabzhenie Petrograda," *Elektrichestvo*, 1924, no. 4: 177; Kotomin, "Deiatelnost," 327.

Section for Communal Economy operated the other three utilities until 3 December 1920, when the ETO took command of the 93 MW of all four stations and established the Unified State Electric Stations—despite opposition from local government. The new state section formally became Petrotok under Glavelektro in March 1922, and Petrotok became Elektrotok when Petrograd became Leningrad in 1924.[59]

Geography may not be destiny, but access to fuel shapes future options. Leningrad's prewar fuel, Cardiff coal, was politically undesirable, and the poor railroad system aggravated the city's distance from southern oil and coal. Hydropower depended on the completion of the Volkhov station. Thus the fuel situation remained "fully undetermined and disordered" through the mid-1920s.[60] Wood generated two-thirds of the electricity during the darkest days of 1920 and remained a major Petrograd fuel through 1924, when it was quickly eclipsed by the return of Donets coal, as Graph 7.2 illustrates.[61] Once accessible, southern oil and coal supplied three-quarters of Leningrad's fuel. The growing role of peat came from the expansion of the Red October regional station.

Supplies of domestically produced materials and equipment remained precarious. In March 1920, the Red October (formerly Utkina Zavod) project, halted during the war, received only fractions of materials, ranging from no iron to 69 percent of wood. Similar shortages hindered the construction of the Volkhov and connecting substations.[62] These problems limited the Red October station initially to 4 MW in 1923, down from the original 30 MW and revised 10-MW forecasts. The peat-fired station gradually increased capacity to 10 MW in 1925 and 20 MW in 1926, although unpublicized defects kept it from operating fully until late 1927.[63] Equipment shortages also frustrated plans

[59] "Petrogradskaia konferentsiia rabotnikov elektricheskoi promyshlennosti," *Izvestiia Elektrotresta*, May 1920, 3; "Biulleten," *Elektrichestvo*, 1923, no. 2: 118; Elektrotok, *Statisticheskii spravochnik*, 9.

[60] A. A. Kotomin, "O snabzhenii elektricheskoi energiei Petrogradskogo raiona v techenie 3-kh let," in *Elektrifikatsiia Rossii: Trudy 8-ogo Vserossiiskogo elektrotekhnicheskogo sezda v Moskve 1–10 oktiabria 1921* (Moscow: Gosizdat, 1921), 2: 148; Anatolii V. Venediktov, ed., *Vosstanovlenie promyshlennosti Leningrada (1921–1924 gg.)* (Leningrad: Izdatelstvo Leningradskogo universiteta, 1963), 1: 269.

[61] Elektrotok, *Statisticheskii spravochnik*, 32, 48–49.

[62] Kotomin, "O snabzhenii elektricheskoi energiei," 136; Osadchii, *Materialy k XIX sessii plenuma TsESa*, 6, 10; S. I. Ikonnikov, *Sozdanie i deiatelnosti obedinennykh organov TsKK-RPI v 1923–1924 gg.* (Moscow: Nauka, 1971), 333.

[63] Vulf, "Elektrosnabzhenie Petrograda," 10; "Khronika," *Elektrichestvo*, 1925, no. 12: 743; *Obzor sostoianiia rabot*, 6; "Biulleten," *Elektrichestvo*, 1927, no. 1: 43; Elektrotok, *Statisticheskii spravochnik*, 11.

Graph 7.2. Leningrad fuel use, 1920–27

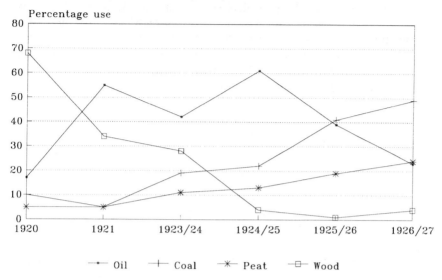

Source: Elektrotok, *Statisticheskii spravochnik 'Elektrotoka' 1913–1928* (Leningrad, 1929), 32, 48–49.

to harness surplus factory capacity. Of three factories, only Obukhov, operating on the same frequency and phase as the 1886 Company station, produced electricity for Petrotok in 1921. The lack of transformers and cables delayed the conversion of Putilov until 1923 and ended any chance to use Metall.[64] On the demand side, higher tariffs and the installation of 40,000 electric meters reduced the lighting load to prerevolutionary levels by 1924.[65]

Finances continually troubled Petrotok. Before the reestablishment of tariffs, free electricity and the kustar heaters for lightbulb sockets resulted in high consumption despite the lack of commercial and industrial use and street lighting.[66] As part of the NEP's focus on economic feasibility, Petrotok had to pay for fuel after May 1922, placing the utility in a desperate financial situation. Delayed payment from some users and nonpayment by others, including the state govern-

[64] Kotomin, "O snabzhenii elektricheskoi energiei," 137–38; Vulf, "Elektrosnabzhenie Petrograda," 9.
[65] Kotomin, "Deiatelnost," 328. This increased Leningrad's electric meters by one-half.
[66] Vulf, "Elektrosnabzhenie Petrograda," 7.

ment, exacerbated the utility's problem. One response was to inte-
grate vertically by taking control of its peat supply from its government suppliers, decreasing Elektrotok's cost by a third and increasing
availability.[67]

By mid-decade, the question of future fuels remained open, but the
debate had shifted from availability to the economics of investment.
Major industrial customers did not rebuild their stations, economically more advantageous than paying Elektrotok tariffs, only because
they lacked the funding and state planning organs pressured them to
wait for the Volkhov hydrostation.[68]

Delays in completing the Volkhov station, problems with the Red
October, and NEP-boosted industrial demand caused serious shortages in 1925–26. Elektrotok prohibited new hookups, and some factories reopened their own stations in response to cutbacks.[69] Until the
Volkhov station opened, Leningrad endured the worst electric supply
situation of the first-tier cities, a victim of its prerevolutionary inability
to act on hydropower. Although the Volkhov's 54 MW nearly increased Leningrad's capacity by half, rising industrial demand ensured that the fuels of the future would burn as well as bubble.

Despite the near collapse of the city's power supply in 1920–21,
Moscow recovered quicker than Leningrad. The MOGES consisted on
1 January 1922 of eight stations with 108 MW of installed capacity, 81
MW of which were 50-cycle 3-phase AC, the industrial standard, and
23 MW of which were 25-cycle 3-phase AC from the tram station.[70]
The 1st Moscow State Electric Station supplied half the capacity, and
Elektroperedacha and the tram station provided another 40 percent.
Three small stations near Elektroperedacha—Pavlov, Glukhov, and
Orekhov—supplied the other 10 percent via the peat-fired station.

Recovery occurred on three fronts: organization, fuels, and equipment. When the NEP "delivered [Moscow utilities] from the excessive
guardianship of numerous agencies which mixed in its life" and sta-

---

[67] "Elektrotekhnicheskaia promyshlennost," *Ekonomicheskaia zhizn*, 11 May 1922; Kotomin, "Deiatelnost," 331.

[68] E. g., 1 kWh cost the Red Vyboretz (formerly Rozenkrantz) factory 7.3 kopecks
versus the 9.0 charged by Elektrotok; see A. Gorev, "Planovaia elektrifikatsiia," *Pravda*,
4 January 1925, 1; see also A. A. Gorev, "Elektrifikatsiia SSSR," *Planovoe khoziaistvo*,
1925, no. 2: 171, 173.

[69] Elektrotok, *Statisticheskii spravochnik*, 11; A. A. Kotomin and M. D. Kamenetskii,
"Obzor deiatelnosti Leningradskogo obedineniia gosudarstvennykh elektricheskikh
stantsii 'Elektrotok' za period 1917–1927 gg.," in "Izvestiia Elektrotoka," *Elektrichestvo*,
1928, nos. 1–2: 4.

[70] V. I. Ianovitskii, "Elektrosnabzhenie Moskvy i blizhaishie perspektivy v etoi oblasti," *Elektrichestvo*, 1922, no. 1: 21–22.

tion operations concentrated in the MOGES, reconstruction began in earnest.[71] The arrival of spare parts in 1922 greatly improved output, although working turbine capacity still exceeded boiler capacity by 20 MW (20 percent) in 1923.[72] The revival of oil and peat supplies essentially eliminated wood consumption by 1922, two years before Leningrad.[73] Using specific fuel consumption as an indicator, the MOGES reached its nadir in 1919–20 and regained prewar levels of efficiency in 1924–25 with the completion of capital repairs. Industrial consumption reached its prewar share of two-thirds of output in 1925–26.[74]

Consumption expanded rapidly at an "American tempo" due to the NEP-prompted industrial recovery, forcing an expansion of capacity at the 1st Moscow and Elektroperedacha stations.[75] The MOGES matched 1916 levels in 1922–23. Three years later, its output of 498 MkWh doubled the 1916 level and represented 45 percent of first-tier and 35 percent of all utility output. Most of the increase came from three regional stations that accounted for half (75 MW) of the MOGES's 151 MW and 60 percent of output in 1925–26 (see Graph 7.3).[76]

Glavelektro, Elektroplan, and the Moscow regional planning commission developed two five-year MOGES plans, one in 1922 for 1923–27 and a second in 1925 for 1925–30. As with Leningrad, the plans seriously underestimated demand and the MOGES's ability to meet it.[77] The major change between the two plans was the perception of fuels. The 1923–27 plan viewed oil as a valuable commodity for export and the 1st Moscow Station as a peaking and reserve plant. The 1925–30 plan envisaged doubling the capacity of existing stations from 151

[71] Robert E. Klasson, TsGANKh f. 9508, op. 1, ed. kh. 14, 4.

[72] Ianovitskii, "Elektrosnabzhenie Moskvy," 22; K. Lovin, "Kratkii predvaritelnyi otchet o deiatelnosti 'MOGES' za 1923–24 operatsionnyi god," *Elektrichestvo*, 1924, no. 11: 566.

[73] "Khronika kommunalnoi tekhniki," *Kommunalnoe delo*, 1925, nos. 11–12, 88.

[74] "Biulleten," *Elektrichestvo*, 1924, no. 1: 36; 1925, no. 8: 490; S. P. Stafrin, "Rabota elektrostantsii Moskovskogo raiona za 1922," *Elektrichestvo*, 1924, no. 1: 32; "Biulleten," *Elektrichestvo*, 1925, no. 8: 489; 1927, no. 4: 148.

[75] G. M. Krzhizhanovskii, "Elektrifikatsiia nakanune rekonstruktsionnoi polosy nashego khoziaistva," *Ekonomicheskaia zhizn*, 21 January 1926, 2; Ianovitskii, "Elektrosnabzhenie Moskvy," 23–24.

[76] "Biulleten," *Elektrichestvo*, 1927, no. 1: 44; 1928, nos. 1–2: 34–35.

[77] Glavelektro, *Elektrosnabzhenie Moskovskogo raiona na blizhaishee piatiletie (1923–1927 gg.)* (Moscow: Glavelektro, 1922), 3; K. P. Lovin, "Blizhaishie perspektivy elektrosnabzhenii Moskvy," *Elektrichestvo*, 1925, no. 7: 392–93; "Po Moskve," *Izvestiia*, 15 July 1925, 5; 16 July, 6; "Khronika," *Elektrichestvo*, 1922, no. 2: 57; Lovin, "Kratkii predvaritelnyi otchet," 564.

Graph 7.3. Moscow electric output, 1913–26

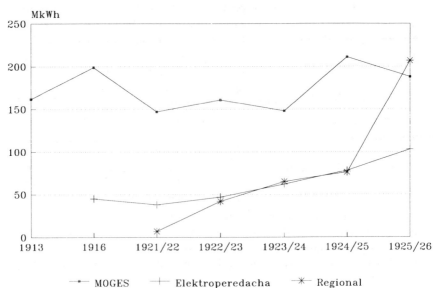

Source: 1913–26: "Biulleten," *Elektrichestvo*, 1927, no. 1: 44.

to 277 MW. Another 90 MW would come from industrial stations or a new large station, fueled by peat, Donets coal, or even Baku oil. By 1930, local fuels would power 148 MW (53 percent) of capacity, compared with 49 percent in 1925; that is, oil-fired capacity would double too, weakening one of the main underpinnings of GOELRO, emphasis on local fuels. As in Leningrad, high-quality and local fuels remained essential in electrifying Moscow.

The shift in perspective from the 1920 focus on local low-quality fuels to 1925, when planners seriously considered the construction of a 50–100-MW oil-fired plant, was significant. A major change from GOELRO's reliance on local fuels, it marked the end of recovery for the MOGES and the start of an electrified future based on a secure economic foundation. As the coal industry and transportation system rebuilt, the virtues of high-quality fuel looked increasingly attractive, precisely the situation Ramzin warned about in 1920.

Moscow realized the promise of GOELRO's emphasis on regional stations. Because it was the capital, access to resources proved easier there than for other cities. More important, its thermal regional stations did not demand the long time and large costs of Leningrad's Volkhov hydrostation. The first postwar regional stations, the temporary 5-MW Shatura and 12-MW Kashira, began operations in 1920 and 1922, respectively, outside Moscow. Despite the large publicity, their initial output proved modest: they contributed only 7 MkWh (4 percent) of Moscow's power in 1921–22. By 1924, however, the two stations contributed 65 MkWh (20 percent) of Moscow's electricity and doubled that percentage and tripled output to 207 MkWh in 1926 after the main Shatura station opened.[78]

Kashira and Shatura were bold steps into the unknown which reflected Moscow's desperate need for new sources of electricity, and they suffered accordingly. They began as patchwork stations, "made up of the strangest assortment of parts, collected from all over Russia," and served as testbeds to determine the feasibility of blending Soviet and foreign technology to burn peat and Moscow brown coal.[79] The lack of new Western technology and lack of participation by foreign firms hindered development of both stations.[80] Unlike Elektroperedacha, Kashira and Shatura initially began without direct Western support. GOELRO took this act of independence, of designing and building new types of electric stations, reluctantly. Once Western assistance became available, it was used.

Expanding on Elektroperedacha, Shatura demonstrated the successful large-scale harvesting, preparation, and burning of peat. Opening with 16 MW in December 1925, it doubled quickly to 32 MW and added a third 16-MW Czech turbogenerator in 1927.[81] Shatura was an impressive demonstration of Russian ingenuity and effective cooperation with foreign firms. The basic equipment was imported, but the research and modifications to burn peat efficiently were domestic.

The TsES started planning Kashira in 1918, but construction did not begin until 1920. Although the station had been planned to open with

[78] Klasson, TsGANKh f. 9508, op. 1, ed. kh. 14, 6; "Biulleten," *Elektrichestvo*, 1928, nos. 1–2: 35.

[79] V. Khudiakov, "Gosudarstvennaia Kashirskaia raionnaia elektricheskaia stantsiia," *Elektrichestvo*, 1922, no. 2: 16–24; A. Vinter, "Torfosnabzhenie na raionnykh elektricheskikh stantsiiakh," *Elektrichestvo*, 1922, no. 2: 27–30; P. I. Voevodin, "Otkrytie Shaturskoi GES imeni V. I. Lenina," *Elektrichestvo*, 1926, no. 1: 50–54; Iu. N. Flakserman, *Teplofikatsiia, 1921–1980* (Moscow: Nauka, 1985), 20–24.

[80] R. Ferman, "Elektricheskoe oborudovanie gosudarstvennoi Kashirskoi raionnoi stantsii i elektroperedacha v Moskvu," *Elektrichestvo*, 1922, no. 2: 24.

[81] *Obzor sostoianiia rabot*, 22; "Biulleten," *Elektrichestvo*, 1927, no. 1: 4; K. P. Lovin, "10 let raboty Shaturskoi GRES im. V. I. Lenina," *Elektricheskaia stantsiia*, 1936, no. 1: 35.

the 9th Congress of Soviets in December 1921, the lack of skilled personnel to install the transformers delayed initial operations until July 1922. The installers ultimately came from Sweden. Two 6-MW Brown-Boveri turbogenerators, powered by brown coal, sent electricity 130 kilometers over a 115-kV transmission line.[82]

As Moscow became more dependent on high-voltage transmission lines, their reliability became an issue. From 1924 to mid-1925, the lines serving Moscow failed seventy-two times, an average of once a week. These outages, which rarely lasted less than an hour, harmed electrical equipment and caused financial and material losses to users. Equally important, they diminished confidence in the MOGES and strengthened factory interest in independent stations. Short circuits resulted from technical problems, including material fatigue, overloading, and poor equipment, but carelessness, hooliganism, and natural disasters also caused many interruptions. Vandalism, apparently mindless, became a serious problem, with accidental electrocution the only deterrent. Glavelektro responded with research on better equipment and propaganda campaigns against the axes and rifles of vandals.[83]

The first-tier stations recovered and expanded rapidly. By 1924, Moscow and Baku had doubled prewar output. The laggard was Leningrad, which reached that level only three years later, after the Volkhov hydrostation opened. The former capital had lost its technological leadership to Moscow, a shift paralleling the overall transfer of political and economic power. In a graphic demonstration of the long-term consequences of technological decisions, the ex-capital's earlier inability to develop hydropower cost it dearly, whereas the 1886 Company's prerevolutionary investment in peat outside Moscow paid handsome dividends. The decision to concentrate Leningrad's future electricity on the previously untried hydropower meant that the overruns of schedule and budget standard in the introduction of any new technology further crippled the city's recovery. In all three cities, supply failed to meet demand, but, with the exception of Leningrad, this failure stemmed more from the very rapid increase in industrial consumption than from inability to expand output. Utilities elsewhere

[82] I. G. Grishkov, "Kashirskaia elektrostantsiia—perevenets Leninskogo plana elektrifikatsii," *Istoriia SSSR*, 1960, no. 2: 217–23; Charles P. Steinmetz, "Russia's First Regional Power Station," *Electrical World*, 25 November 1922, 1155.

[83] A. Eisman, "V chem zalog uspekha elektrifikatsii?" *Izvestiia*, 18 July 1925, 3; L. Trotsky, "Budem okhraniat elektroprovody," *Izvestiia*, 18 July 1925, 3; R. Klasson, "Budem okhraniat elektroprovody," *Izvestiia*, 26 July 1925, 1; "Po Moskve," *Izvestiia*, 10 July 1925, 5, and 14 July, 4.

faced similar problems but without the same financial and administrative resources.

Cities

URBAN ELECTRIFICATION occurred within serious financial, material, and organizational constraints. Like other enterprises, utilities under the NEP faced a difficult enviroment very different from war communism. The foremost change, dictated from Moscow, was economic. Charges were reinstated for city services as enterprises moved from state subsidies to the economic profitability of *khozraschet* (*khoziaistvennyi raschet*), financial self-suffiency.[84] In many ways, utilities were in a better condition than city services such as housing, which suffered actual destruction. The basic plant of utilities remained intact, if run down. The estimated value of municipal property dropped from 2.7 to 2.1 billion rubles between October 1917 and October 1923. The major losses came in streets and transportation (1.7 to 1.2 billion rubles); the value of utilities, however, dropped only marginally (284 to 268 million rubles).[85] Nonetheless, local stations, returned to a devastated economy with a mandate for economic viability, were "close to catastrophe."[86]

Electric light, power, and traction recovered quicker than the rest of the economy because of their less decrepit state but also thanks to a state commitment of resources, however inadequate that seemed to the managers on the spot. Utilities surpassed 1916 output in 1924–25, and the number of operating tram systems climbed from twenty in 1920 to thirty-eight in 1924.[87] Theoretical capacity remained substantially greater than actual output through 1926 as unrepaired equipment and lack of key components such as transformers limited generation.

Two trends characterized the growth of utilities during these years.

[84] "O vzimanii platu za uslugy, okazyvaemye predpriiatiiu kommunalnogo khoziaistva," *Kommunalnoe delo*, 1921, no. 1: 48–49; R. W. Davies, *The Development of the Soviet Budgetary System* (Cambridge: Cambridge University Press, 1958), 55–56.

[85] G. Pisarev, "Kommunalnoe khoziaistvo i ego organizatsionnye osnovy," *Kommunalnoe delo*, 1921, no. 1: 16; M. Petrov, "Osnovnoi kapital kommunalnogo blagoustroistva RSFSR," *Kommunalnoe delo*, 1926, no. 23–24: 62. This is twice the 139 million rubles counted by Diakin; see V. A. Diakin, *Germanskie kapitaly v Rossii* (Leningrad: Nauka, 1971), 268–69. Different accounting bases and ruble values may cause the difference.

[86] Kuibyshev, "Sostoianie elektrotekhnicheskoi promyshlennosti," 41.

[87] "Biulleten," *Elektrichestvo*, 1927, no. 1: 43; "Electric Trolleys in the U.S.S.R.," *Russian Review*, 1 June 1925, 228.

Table 7.2. Size of electric stations, 1923

Size (kW)	Number[a]		Average capacity (kW)[a]	
0–50	641	(60)	21	(2)
51–200	224	(21)	109	(5)
201–500	86	(8)	326	(5)
501–1000	38	(4)	790	(6)
1000–2000	39	(4)	1,436	(11)
2001–5000	28	(3)	2,929	(16)
subtotal	1,056	(99)	205	(45)
5000+	15	(1)	19,400	(55)
total	1,071	(100)	526,000	

Source: "Biulleten," *Elektrichestvo,* 1923, nos. 7–8: 401.
[a] Percentage of total in parentheses.

First-tier capacity and output increased more quickly than second- and third-tier capacity, widening the gap between the first and other tiers (see Tables 7.2 and 7.3). Moscow, Leningrad, and Baku accounted for 60 percent of installed capacity and 78 percent of produced power in 1926, compared with 55 percent and 62 percent, respectively, in 1913.[88] While first-tier output grew from 431 MkWh in 1913 to 866 MkWh in 1924–25 and 1,121 in 1925–26, second- and third-tier output went from 259 to 266 to 314 MkWh. A second trend was the great increase in the number of utilities as small towns and villages extended power lines from factories, converted mills, and built new stations. This increase in numbers sharply decreased station size in a continuation of prewar trends (see Graph 3.3). Urban utilities nearly tripled to 640 between 1913 and 1926, although installed capacity in the second- and third-tiers increased only by half from 151 to 224 MW. Consequently, average capacity dropped from 683 kW in 1913 to 455 kW in 1920 and to 350 kW in 1926, a sign of the geographic diffusion. Rural stations, a separate category, had an average capacity below 20 kW, a factor of twenty smaller.[89] A broader 1923 survey that included rural stations found that 641 of 1,071 stations (60 percent) were less than 50 kW and had 2 percent of total capacity.[90] The 90 percent under 500 kW had only 12 percent of total capacity, and the 1 percent of stations over 5 MW had 55 percent.

[88] "Biulleten," *Elektrichestvo,* 1927, no. 1: 43; 1928, nos. 1–2: 35.

[89] "Biulleten," *Elektrichestvo,* 1927, no. 1: 43.

[90] "Biulleten," *Elektrichestvo,* 1923, nos. 7–8: 401–3. The survey also included seventy-three nonreporting stations.

Table 7.3. Capacity and production, October–December 1924

City	Capacity (MW)[a]		Output (MkWh)[a]		Per capita kWh
Moscow	116	(26)	92	(32)	217
Leningrad	101	(23)	53	(18)	212
Baku	69	(16)	71	(25)	952
subtotal	286	(65)	216	(75)	
Kiev	19	(4)	9	(3)	83
Odessa	17	(4)	6	(2)	74
Rostov-on-Don	8	(2)	4	(1)	70
Kharkov	7	(2)	6	(2)	66
Ekaterinoslav	5	(1)	2	(1)	96
Other	97	(22)	44	(15)	32
subtotal	153	(35)	71	(25)	
total	439	(100)	287	(100)	118

Source: "Biulleten," *Elektrichestvo*, 1925, no. 3: 192.
[a] Percentage of total in parentheses.

The prerevolutionary preference for DC over AC continued despite Glavelektro's efforts to impose AC on new and expanded stations. The advantages remained the same: compared with AC's distant benefits, DC offered municipalities immediate, firm advantages, including lower construction costs and a larger base of DC knowledge and equipment.[91] Existing utilities usually rebuilt with the same current to avoid the economic and technical costs of AC conversion, although some cities converted from 1- to 3-phase AC.[92] Only a few switched from DC to AC.[93]

Although industrial load remained concentrated in a few cities, the absolute and relative levels of utility-based industrial consumption increased greatly. This increase grew in large part from the 131 million rubles invested in industrial electrification through 1926.[94] More than 90 percent of sixty-three large utilities carried some industrial load by 1925–26. Of twenty-two second-tier cities for which comparative data

[91] P. Skvortsov, "Novye zadachi elektrosnabzheniia," *Kommunalnoe delo*, 1924, no. 6: 21–22.
[92] E. g., Irkutsk and Orenburg; "Kommunalnye predpriiatiia," *Kommunalnoe delo*, 1925, no. 6: 89; no. 1: 34.
[93] E. g., Semipalatinsk intended to rebuild with a higher voltage DC system, but the GUKKh advised the city to reequip with 3-phase AC to meet future loads; "Na mestakh," *Kommunalnoe delo*, 1924, no. 1: 73.
[94] Elektrobank, *Finansirovanie elektrokhoziaistva (dva goda raboty Elektrobanka)* (Moscow-Leningrad, 1927), 10–12, 40.

exist, the industrial load more than doubled in half, increased by a lesser degree in a sixth, and remained the same or decreased in a third compared with 1914. Nonetheless, lighting far outweighed industrial usage in forty-five cities, equaled it in seven, and trailed in only eleven.[95] These eleven cities, however, included five of the seven cities with a capacity over 10 MW[96] and four of the thirteen cities between 2.5 and 10 MW.[97] Only two of forty-three utilities with a capacity less than 2.5 MW had heavy industrial loads.[98] The promise of industry rationalized, mechanized, and modernized by electrification held true primarily for the larger cities, but the utilities of the second tier had progressed toward this goal, albeit from a very small prewar base.

The NEP reopened the important political question of who should control utilities. Before the NEP, Glavelektro controlled urban utilities. The NEP placed Glavelektro and Gosplan on the defensive; their pleas that the GOELRO plan depended on continued direct control of all utilities lost to the cities' claims that the utilities were an integral part of the urban economy and, with the need to balance budgets, of city coffers. Although cities feared Glavelektro's political strength because of Lenin's support, their economic argument eventually triumphed, but not without a fight.[99]

The first round went to the cities. The communal section (kommunalnyi otdel) of an *ispolkom* (*ispolnitelnyi komitet*, executive committee) provided urban government. Under an 8 April 1920 SNK decree, communal sections controlled all enterprises of "general use with local significance," including lighting, water, and transportation.[100] Enterprises changed from autonomous entities to municipal enterprises under a local authority. The NKVD's GUKKh provided general control and a conduit to the central government. As utilities returned to municipal control, the GUKKh became increasingly important as their representative in the state government.[101]

On 15 April 1920, the VSNKh petitioned "to militarize all state elec-

[95] "Statisticheskie svedeniia . . . za 1914 god," *Elektrichestvo*, 1917, nos. 4–6: 98–103; "Biulleten," *Elektrichestvo*, 1927, no. 4: 148.

[96] Baku, Leningrad, Moscow, Nizhni-Novgorod, and Odessa. The exceptions were Kiev and Kharkov.

[97] Gubakha (Kizel), Ivanovo-Voznesensk, Kremenchug, and Tula.

[98] Ginduksh and Cheliabinsk.

[99] M. Zemblukhter, "V. I. Lenin i kommunalnoe khoziaistvo," *Kommunalnoe delo*, 1924, nos. 1–2: 7.

[100] And less obvious enterprises like laundries and barbershops; see "Zakonodatelnyi otdel," *Kommunalnoe khoziaistvo*, 1921, nos. 1–2: 31.

[101] V. Levi, "Mestnye sezdy po elektrokhoziaistvu," *Kommunalnoe delo*, 1925, no. 1: 12.

tric stations under a unified central administration and to consider them enterprises of state importance."[102] Although independent of GOELRO, this proposal meshed with the logic of war communism and the electrical engineers' efforts to create an all-powerful main electrotechnical committee to implement the GOELRO plan. On 7 June 1921, the ETO won possession of city stations from the notorious NKVD communal sections.[103] The communal sections, upset by this failure, counterattacked.[104] Far more than the promise of rural electrification at the 8th All-Russian Electrotechnical Congress, this effort was a decentralized, broad-based political movement motivated by economic self-interest and self-definition. An article in the GUKKh journal *Kommunalnoe Delo* declared the transfers "senseless and done only under the influence of the beautiful, celebrated slogan, 'the electrification of Russia,' a slogan impractical due to a complete absence of materials and machines. . . . The electrifiers and all those who are so deeply carried away by electrification to shout slogans in the air [are] tearing flesh from a living body."[105]

The angry communal sections viewed the transfers as part of a larger effort by state organs to control local operations. The KGS tried to monopolize construction, the Commissariat of Land tried to control suburban lands and gardens, and the Commissariat of Health tried to direct city activities ranging from sewage to baths. "But the most persistent attack against one of the best parts of the city economy is conducted by Glavelektro."[106] Glavelektro control led to unusual divisions of responsibility as well as bad feelings. In Nizhni-Novgorod, Glavelektro ran the combined electric and water station, while the communal section operated the distribution network. In Samara and Smolensk, "under the slogan of electrification, [Council for the Econ-

[102] "Deiatelnost prezidiuma VSNKh," *Narodnoe khoziaistvo*, 1920, nos. 9–10: 33. The verb "militarize" takes on particular meaning because Trotsky, acting commissar of transportation, proposed the militarization of labor to advance economic recovery at the 1920 8th Congress of Soviets that approved the GOELRO plan; see Segal, *Leon Trotsky*, 244–46.
[103] V. Levi, "4-i sezd Gubelektrootdelov," *Elektrichestvo*, 1922, no. 1: 50. His interpretation should be regarded as somewhat biased, though not necessarily inaccurate. Levi changed from an antagonist to an advocate of municipal control and became the main writer on electrification for the GUKKh's *Kommunalnoe delo*.
[104] Resolution of the Third All-Russian Congress of Professional Unions of City Workers, "Sezdy i konferentsiia," *Kommunalnoe delo*, 1921, no. 1: 93.
[105] Pisarev, "Kommunalnoe khoziaistvo," 17.
[106] "Organizatsionnye voprosy," *Kommunalnyi rabotnik* 72–73 (1921): 16; A. Brauner, "K voprosu o peredache elektricheskikh stantsii obshchemu polzovaniiu elektrootdelam," *Kommunalnoe delo*, 1921, no. 1: 36.

omy] electric sections have taken both the electric station and tram (what does a tram have to do with electrification?)."[107]

Like proponents of small-scale rural stations, cities opposed not electrification, which would transform the country "into a second North America," but central control by Glavelektro, which "terrorized" local authorities by depriving them of materials and equipment.[108] Pride played a role too: after keeping them operating through the civil war, why should the cities lose their utilities now? More important, the NEP demanded the retention of the profitable electric stations to subsidize other city enterprises—exactly what Glavelektro feared. A lesser factor was trained personnel: unless the staff were transferred with the station, city advocates claimed, Glavelektro lacked the people to operate the stations.[109]

In November 1921, the VSNKh placed the utilities in a region under a *gubelektrootdel* (regional electric section), assuming that a unified administration of physically separated powerplants served the state and society better than local control.[110] The cities disagreed. In February 1922, Gosplan, Glavelektro, and the GUKKh, responding to pressure from the All-Russian Central Executive Committee and the SNK, established two categories of utilities.[111] Glavelektro retained control of all large, regional, and unified stations. All other stations fell under local control. Glavelektro then negated any amiable resolution by assigning ninety-three urban utilities to the first category; that is, every major city remained under its aegis. The NKVD again petitioned the SNK to return utilities to their cities, albeit with Gosplan oversight and limited joint management by local Glavelektro and GUKKh organs. The NKVD also suggested turning utilities into joint stock companies. The GUKKh protested its parent organization's proposal and demanded single, not joint, management while emphatically opposing private companies. Utilities needed private capital, but private ownership would negate municipal control, a heartfelt prerevolutionary cause of many city officials and electrical engineers.[112] On 30 June 1922, the Council for Labor and Defense ordered the transfer of

---

[107] "Khronika Tsentra," *Kommunalnoe delo*, 1924, no. 6: 55; Pisarev, "Kommunalnoe khoziaistvo," 17.

[108] Brauner, "K voprosu," 37–38.

[109] "Organizatsionnye voprosy," *Kommunalnyi rabotnik* 72–73 (1921): 16. Staff did transfer—and municipalities did not like losing skilled personnel.

[110] "Novaia forma ekspluatatsii elektricheskikh stantsii," and "Polozhenie o gubernskikh pravleniiakh elektrostantsii obshchestvennogo polzovaniia," *Ekonomicheskaia zhizn*, 25 November 1921, 1, and 28 November 1921, 1.

[111] "Iz zhizni," *Elektrichestvo*, 1922, no. 2: 55–56; M. Zemblukhter, "V. I. Lenin i kommunalnoe khoziaistvo," *Kommunalnoe delo*, 1924, nos. 1–2, 7.

[112] "Gorodskie elektricheskie stantsii," *Kommunalnoe khoziaistvo*, 1922, no. 7: 10–11.

twenty-three city utilities from Glavelektro to local control.[113] Future transfers eventually deprived Glavelektro of all but the Moscow, Leningrad, and Baku utilities. Cities now directly controlled the operations and finances of their utilities. The struggle to guide, if not control, Soviet utilities continued, albeit in several different arenas, none overtly but all quite political.

A fierce struggle erupted in 1923 over station construction and operation among GOELRO, Gosplan, GUKKh, and a new party, the Ukrainian government, which sought to exclude its utilities from central control. As part of its efforts to maintain its economic independence, the Ukrainian government supported the GUKKh's claims and stated that only stations having all-union significance should need Gosplan's permission.[114] This struggle for control of new stations continued until late 1924, when, after further SNK-prompted negotiations, Gosplan, the GUKKh, and Glavelektro agreed on four levels of authorization: stations below 50 kW now required only preliminary permission from local government; stations of 50–500 kW required the approval of the regional authorities; stations of 500 kW–3 MW needed approval from the republic VSNKh and NKVD; and stations above 3 MW were the province of Glavelektro.[115] In a defeat for Glavelektro and a sign of increasing municipal aggressiveness, the agreement demarcated clear lines of authority among local, regional, republic, and national organs. The small capacities of the non-Glavelektro stations indicated the low levels of present and predicted generation outside the first tier. Demarcation, however, did not end the war for control; Glavelektro and Gosplan tried to expand their influence on utility operations. Glavelektro proposed, for example, to maximize the use of trained personnel by basing them at the regional level instead of individual utilities to respond to the needs of a larger area. Municipalities viewed such efforts at "rationalization" as oppressive financial and administrative burdens designed to increase central power.[116]

---

[113] "Ofitsialnyi otdel," *Kommunalnoe delo*, 1922, no. 2: 111.

[114] "Khronika Tsentra," *Kommunalnoe delo*, 1924, no. 6: 50; E. H. Carr and R. W. Davies, *Foundations of a Planned Economy, 1926–1929*, vol. 1, pt. 1 (New York: Macmillan, 1969), 362–63.

[115] "Khronika Tsentra," *Kommunalnoe delo*, 1924, nos. 11–12: 54; "O poriadke sooruzheniia i registratsii elektricheskikh stantsii i nadzora za takovumi," and "O poriadke soglasovaniia mestnogo elektrostroitelstva s obshchimi i mestnymi planami elektrifikatsii," *Sbornik dekretov, postanovlenii, rasporiazhenii i prikazov po narodnomu khoziaistvu*, 1925, no. 43: 153; no. 44: 174.

[116] Iu. Mitlianskii, "Organizatsionnye voprosy elektricheskikh stantsii," *Kommunalnoe delo*, 1924, no. 6: 12; "Ofitsialnyi otdel," *Kommunalnoe delo*, 1924, no. 9: 37; "Khronika Tsentra," *Kommunalnoe delo*, 1924, nos. 11–12: 54; "Kommunalnye predpriiatiia," *Kom-*

Responding to Glavelektro's proposals, the 1925 All-Russian Conference of Communal Electric Station Managers proposed a GUKKh bureau to manage utility activities. The bureau's significance was that the GUKKh would perform activities desired by Glavelektro. The issue, again, was control. The GUKKh did create the bureau, but it played only a minor role, which may have been its sponsors' intention.[117]

Securing local control was only one issue facing utilities. Others included repairing and expanding existing stations, constructing new stations, and switching to self-supporting finances. As with the first tier, the immediate problem was the deplorable state of present utilities. In 1922, two-thirds of operable Diesel engines needed capital repairs, as did half the working steam engines. Half the country's turbogenerators were out of service and the other half overworked.[118]

The restoration of the country's devastated financial structures helped reestablish utility finances.[119] The establishment of a new banking system aided the gradual return of the economy to monetary from commodity exchange.[120] Nonetheless, financial solvency became a major problem as utilities had to fund repairs and expansion, guard their income from other city enterprises, and balance operating budgets.[121]

The introduction of khozraschet and the resumption of tariffs posed two major questions for Glavelektro, the utilities, and the VSNKh. First, it was unclear which indirect costs, particularly amortization, tariffs should cover, and concurrently profit had to be defined and

---

munalnoe delo, 1925, no. 5: 60–61; "V Gosplane RSFSR," *Kommunalnoe delo*, 1926, no. 5: 52.

[117] "Vserossiiskoe soveshchanie zaveduiushchikh kommunalnymi elektrostantsiiami," *Kommunalnoe delo*, 1925, nos. 15–16: 4; V. Levi, "Ob organizatsii Postoiannogo buro soveshchanii po kommunalnym elektrostantsiiam," *Kommunalnoe delo*, 1925, no. 18: 11–12; "Kommunalnye predpriiatiia," *Kommunalnoe delo*, 1926, nos. 21–22: 101–2.

[118] "Elektrotekhnicheskaia promyshlennost," *Ekonomicheskaia zhizn*, 10 June 1922, 7 May 1922.

[119] Ruble stabilization dramatically decreased budgets. In January 1922, the Moscow utilities had 30 billion rubles in cash, which covered only wages; see "Economic Notes," *Russian Information and Review* 1, no. 13 (1922): 275. By contrast, the 1923–27 plan estimated expenditures of 6.5 billion rubles; see Glavelektro, *Elektrosnabzhenie Moskovskogo*, 15.

[120] As an example of commodity exchange, Volodga acquired all the equipment for a station, save the turbines, from the Academy of Art for 215 kilograms of meat in 1922; "Kommunalnye predpriiatiia," *Kommunalnoe delo*, 1925, no. 5: 61.

[121] Including paying workers. The staff of a Georgian station threatened a strike in 1923 when they were not paid for three months; "Tarifno-ekonomicheskaia rabota Gubotdelov," *Kommunalnyi rabotnik* 89 (1923): 26.

justified.[122] Then there were questions of differing economic and political priorities: should tariffs support widespread electrification or fund local needs? Glavelektro wanted low rates to advance electrification. The agency feared, quite justifiably, that the local authorities would use their utilities to subsidize other municipal operations. Glavelektro favored khozraschet but not capitalist operations; that is, tariffs should cover costs but not profit. Municipal governments wanted higher rates to balance their budgets, and the VSNKh presidium wanted higher rates to finance utility needs and reduce the drain on government coffers.[123]

Questions of accounting for a mixed capitalist-socialist economy did not lend themselves to obvious solutions, yet utilities needed immediate answers. What was profit? Should tariffs be based on class or economic principles? What was the appropriate level of amortization? The answers had serious consequences and sparked long, heated debates.[124] A class principle guided the reimposition of tariffs in 1921–22 with different rates for workers, bourgeoisie, and other users. By 1925, utilities set tariffs to cover costs, increase efficiency, and maximize load. A common definition of costs included amortization, funds for repair, and a 10 percent profit on capital.[125]

Another question without easy answers was the appropriate degree of utility integration with other city enterprises. Integration promised savings from shared facilities, such as repair shops, and a higher load factor from electrifying city enterprises.[126] Administratively, unification promised decreased overhead and more rational use of limited financial and material means.[127] The extent and benefits of unification, however, depended greatly on local conditions.[128] Combining too

---

[122] I. A. Skavani, "K voprosu o sebestoimosti i tarifikatsii energii na russkikh tsentralnykh elektrostantsiiakh," *Elektrichestvo*, 1924, no. 4: 195.

[123] V. L. Levi, "Novaia politika v dele ekspluatatsii elektrostantsii," TsGANKh f. 5208, op. 1, ed. kh. 69, 56; "Khronika," *Elektrichestvo*, 1923, no. 10: 531; no. 11: 596.

[124] E. g., F. Dits, "K voprosu ob ustanovlenii sebestoimosti i tarifov na energiiu, otpuskaemuiu kommunalnym elektro-stantsiiam," *Kommunalnoe delo*, 1925, no. 20: 20–31.

[125] V. Levi, "Pervoe Vserossiiskoe soveshchanie zaveduiushchikh kommunalnymi elektrostantsiiami," *Kommunalnoe delo*, 1926, no. 1: 48.

[126] Water supply was of particular interest because electricity powered only one-third of the country's 325 stations; see N. Bragintsev, "Elektrifikatsiia vodoprovodov," *Kommunalnoe delo*, 1925, no. 4: 7–16.

[127] F. Dits, "Organizatsiia upravleniia kommunalnym elektro-stantsiiam," *Kommunalnoe delo*, 1925, nos. 21–22: 20–25.

[128] E. g., Vologda merged its electric station, water station, baths, and slaughterhouses ("Tresty," *Kommunalnoe delo*, 1925, nos. 15–16: 94), whereas Arkhangelsk only united its electrified enterprises into an electrical trust ("Kommunalnye predpriiatiia," *Kommunalnoe delo*, 1925, no. 20: 53).

many functions in one office could be counterproductive. As Kazan discovered, centralizing all municipal accounting left the utility dependent on an outside office of decidedly low competence.[129]

Another disadvantage of integration was financial. Utility income tempted money-strapped municipal budgets. In 1923–24, utilities generated over two-thirds of the surplus from city enterprises in Moscow gubernia.[130] Similarly, 125 utilities contributed one-third of the 9.5 million ruble surplus for 110 Ukrainian cities in 1923–25, a share exceeded only by the 338 abattoirs.[131] On this issue, Glavelektro and utility managers agreed. In December 1925, the 1st All-Russian Conference of Communal Electric Station Managers declared "intolerable any kind of assignment from [utility] profits to local budgets before the reconstruction and widening of electric stations on a scale such that the stations can serve fully local population and industry."[132]

Just as tariffs incorporated prerevolutionary criteria, so did load factor. Increasing load—and not just output—became a major priority for economical operations. As the 1886 Company had discovered decades earlier, lighting alone was insufficient. The GUKKh urged utilities to increase their load factor by hooking up industry, kustar workshops, and other users and suggested that communal sections arrange favorable loans for potential users to purchase motors and other equipment from the electrotechnical manufacturing trusts.[133] In an extension of GOELRO's minimum program, some utilities constructed transmission lines to unify separate stations.[134] Availability and affordability of transformers and cable determined the feasibility of such networks. Attracting the industrial participation necessary for approval from the planning organs often required financial incentives.[135] The overall rise in industrial load shows that utility-based industrial electrification did diffuse outside the first tier.

Although the Volkhov hydrostation remained the biggest example of an unprofitable short-term investment, some local projects exceeded budgets and then requested more money, usually granted to

[129] "Khronika Tsentra," *Kommunalnoe delo*, 1924, nos. 11–12: 74.

[130] "Kommunalnye predpriiatiia," *Kommunalnoe delo*, 1925, no. 2: 32–34.

[131] "Korrespondentsiia," *Kommunalnoe delo*, 1925, nos. 13–14: 73–75.

[132] Levi, "Pervoe Vserossiiskoe soveshchanie," 42.

[133] "O prisoedinenii k elektricheskoi stantsii motornoi nagruzki," in "Ofitsialnyi otdel," *Kommunalnoe delo*, 1924, no. 9: 37.

[134] E. g., linking the Rostov-on-Don city station to the Artem mines station and the future Shterov regional station; "Kommunalnye predpriiatiia," *Kommunalnoe delo*, 1925, no. 8: 59.

[135] "Kommunalnye predpriiatiia," *Kommunalnoe delo*, 1925, no. 7: 64.

prevent a total loss.[136] Understandably, Gosplan and Elektroplan demanded more financial and operational data from utilities before approving proposals and rejected requests for credit if load estimates seemed suspiciously optimistic.[137]

Financing

BY 1924, UTILITIES had recovered from the tribulations of the previous decade. As second- and third-tier stations started to expand, they encountered a major obstacle in an area familiar to prewar and Western utilities—securing adequate finances. The problem was threefold: lack of funding hindered investment, regional stations consumed the lion's share, and Glavelektro increasingly controlled the flow of funding. Before 1917, utility financing came from a variety of domestic and foreign sources. In the 1920s, funding came from four sources: the state budget, state banks, regional joint stock companies, and local joint stock companies. The differences between 1914 and 1924 were the concentration of financing in central state organs, the absence of foreign investment, and the minor role of private capital. Because it was so limited and controlled, credit played a major role in shaping electrification after 1925 as funding priorities increasingly intertwined with state priorities of planning and foreign trade. Unlike prewar conditions, these financial contraints were overtly political as well as economic.

The hierarchy of tiers was reflected in the emergence of different credit agencies for different strata of utilities. Regional stations received direct state funding and loans from Elektrobank. City and local stations sought credit directly from the state, from Elektrokredit and its successor, Elektrobank, from the Central Communal Bank, and from regional and local joint stock companies.

GOELRO did not discuss the actual mechanics of financing, because of its lack of experience and assumptions of foreign involvement. Realization of the need for specific financial mechanisms appeared slowly. In February 1922, Krzhizhanovskii called for a foreign-funded bank for electrification. European capital would participate because "Russia can be a powerful consumer of European industry [while] the base of European raw materials is moving to the East."[138] Materials and markets: what better enticements for the capitalist world?

---

[136] Ikonnikov, *Sozdanie i deiatelnosti obedinennykh organov TsKK-RPI*, 334.
[137] "Kommunalnye predpriiatiia," *Kommunalnoe delo*, 1925, no. 7: 64–65; no. 8: 57–58.
[138] "Inostrannyi kapital i elektrifikatsiia Rossii," *Trud*, 15 February 1922, 4.

Early in 1922, a Glavelektro commission proposed a state-controlled "Elektrobank," based on German and Japanese models, to channel 60 million rubles of capital from state and public organizations, individuals, and foreign sources.[139] To encourage private investment, this electro-industrial bank would pay dividends of 8 percent. Neither direct foreign credit or equity would be permitted, but the proposed bank would directly import and export goods and materials, bypassing the monopoly of the Commissariat of Foreign Trade.[140]

The establishment of Elektrokredit[141] in December 1922 must have been a disappointment. This was not a bank but a joint stock company initially allowed only 500,000 rubles in capital, which expanded to 2 million rubles within a year and focused on towns and villages.[142] Elektrobank had bowed to Russia's strained financial state in favor of the smaller, less ambitious Elektrokredit. These limited horizons should not obscure the fact that, for the first time, Russia had a financial institution dedicated exclusively to electrification.

Stockholders subscribed to Elektrokredit more slowly than predicted. Its promoters anticipated equal investment between government organs and cooperatives; instead, cooperatives bought only one-third of the initial offering and the October 1923 offering of 1.5 million rubles. Glavelektro, the GUKKh, and Tsentrosoiuz (the central union of cooperatives) held two-thirds of the shares, with Glavelektro alone holding over one-third.[143] City governments and cooperatives subscribed from self-interest: to receive loans and assistance, they had to belong.

Although interest in investing was low, interest in receiving was high. Elektrokredit received more than one hundred requests for funding in its first two months. Quite reasonably, it concentrated on

[139] G. A. Feldman, "Elektrobank," *Voprosy elektrifikatsii,* 1922, nos. 1–2: 66–71. Feldman is better known for the Feldman–Domar model of investment choice for consumption maximization; see Evsey D. Domar, *Essays in the Theory of Economic Growth* (New York: Oxford University Press, 1957), 223–61.

[140] "Finansy. Gosudarstvennyi promyshlennyi bank dlia elektrifikatsii Rossii," *Ekonomicheskaia zhizn,* 26 January 1922, 2; "Finansirovanie elektrifikatsii," ibid., 23 February 1922, 1; "Komissiia po organizatsii finansirovaniia elektrifikatsii," *Voprosy elektrifikatsii,* 1922, nos. 1–2: 164–67.

[141] Officially, the All-Russian Joint-Stock Company for Financing Local Electrification.

[142] A. Kravchenko, "Khronika," *Elektrichestvo,* 1923, no. 3: 54; "Vserossiiskoe aktsionernoe obshchestvo finansirovaniia mestnoi elektrifikatsii," *Elektrifikatsiia,* 1923, no. 1: 26.

[143] Elektrokredit, *Iz praktiki kooperativnoi elektrifikatsii* (Moscow: Tsentrosoiuz, 1923), 16; "Khronika Tsentra," *Kommunalnoe delo,* 1924, no. 6: 49; "Tsentrosoiuz i finansirovanie," *Ekonomicheskaia zhizn,* 10 November 1922, 3; "Iz zhizni 'Elektrokredita'," *Elektrifikatsiia,* 1923, no. 3: 36–37; nos. 5–6: 33; 1924, no. 4: 27–28.

short-term loans promising immediate results, such as connecting a village with a nearby factory station or completing construction halted for lack of funds.[144] In its first year, Elektrokredit made thirty-four loans, ranging from 465 to 126,000 rubles, for a total 546,000 rubles. The loans lasted from six months to three years, limiting the financial attractiveness of many projects. Approximately one-quarter of Elektrokredit's activities involved trade-in-kind, supplying materials and equipment in lieu of money. Its inability to obtain long-term credit and even short-term loans from the Commissariat of Finance and Gosbank, the state bank, greatly hampered Elektrokredit.[145]

These flaws, coupled with a demand greatly exceeding resources and an improving economy, led to Elektrokredit's absorption into the newly created Elektrobank in May 1924. An actual bank, Elektrobank greatly strengthened the financial resources available for electrification (see Table 7.4). At first a small player in a big sea, Elektrobank soon became a big player in a small lake. In 1925, five state banks financed electrification, including the much larger Gosbank and Prombank and the Central Communal Bank.[146] One year later, Elektrobank had reduced the players to three. Its monopolizing efforts followed three lines: establishing zones of demarcation with other state banks, obtaining state funding, and coordinating funding with state plans.

Elektrobank sought to control not only all loans but also the state budget for electrification. By 1927, its disclaimers about not wanting to monopolize, only to help electrify, had evolved into statements on the need to "concentrate credit in one center to coordinate the financial needs of all areas of the electrical economy (basic industries, electrotechnical industry, construction, use), and this center must be Elektrobank."[147] Elektrobank advanced its centralized control, despite early years "burdened by an atmosphere of skepticism and doubt and more than cold relations from our older brothers in the credit area."[148] Agreements with other state banks, the Commissariat of Finance, and regional joint stock firms assuaged these doubts. Elektrobank's main

[144] "Mestnaia elektrifikatsiia v Rossii," *Elektrifikatsiia*, 1923, no. 1: 21; A. K., "Iz deiatelnosti 'Elektrokredita'," in *Iz praktiki kooperativnoi elektrifikatsii*, 10.
[145] "Iz zhizni 'Elektrokredita'," *Elektrifikatsiia*, 1923, no. 2: 37; "Khronika Tsentra," *Kommunalnoe delo*, 1924, no. 6: 48–49.
[146] The first two banks held 83 percent of the 2.6 billion rubles in state banks on 1 September 1925; "Finansy i kredit," *Ekonomicheskaia zhizn*, 6 January 1926, 3.
[147] Elektrobank, *Nekotorye itogi deiatelnosti Elektrobanka* (Moscow: Elektrobank, 1925), 18; Elektrobank, *Finansirovanie elektrokhoziaistva*, 13.
[148] Elektrobank, *Nekotorye itogi deiatelnosti Elektrobanka*, 13.

Table 7.4. Bank credits for electrification, October 1925 and 1926

Bank	1925[a]	1926[a]
Elektrobank	32.4 (55)	64.7 (67)
Gosbank	15.1 (26)	18.8 (20)
Central Communal Bank	1.2 (2)[b]	5.2 (5)
Prombank	7.6 (13)	4.3 (4)
Moscow City Bank	2.2 (4)	4.1 (4)
	58.1	97.1

Sources: Elektrobank, *Finansirovanie elektrokhoziaistva (dva goda raboty Elektrobanka)* (Moscow-Leningrad, 1927), 16. For the Central Communal Bank, see "Mestnye finansy," *Kommunalnoe delo,* 1926, no. 7: 43.
[a] Percentage of total in parentheses.
[b] Elektrobank listed 0.7 million rubles.

potential rival was Prombank, created in 1922 to finance industrial development. From 1923–26, Prombank lent 15.8 million rubles to state industries in electrification, electrotechnology, and films.[149] If it wanted, Prombank could have been a very serious factor in electrification, but other interests beckoned until 1928 when it absorbed Elektrobank. The August 1926 Elektrobank–Prombank agreement eliminated conflict by delineating industrial clienteles and responsibilities, thus reducing the options available to loan seekers. Prombank reduced its lending for electrification, and Elektrobank reduced its lending for nonelectrification activities from 30 percent in November 1924 to 5 percent in October 1926.[150]

Elektrokredit and the Central Communal Bank made decisions without consulting Gosplan and Glavelektro. In contrast, Elektrobank viewed credit as a powerful planning tool and formed tight links with Gosplan, Glavelektro, and regional planning organs to direct the development of electrification.[151] A January 1926 agreement with the

[149] For 1923–24, see *Sev.-Zap. oblastnaia kontora Prombanka k otchetu za 1923–24 god* (Leningrad: Prombank, 1924), 11; for 1924–26, see Elektrobank, *Finansirovanie elektrokhoziaistva,* 16.
[150] Elektrobank, *Finansirovanie elektrokhoziaistva,* 12, 34; and *Nekotorye itogi deiatelnosti Elektrobanka,* 17–18, 45.
[151] A. N. Rumiantsev, "Organizatsiia finansirovaniia melkogo elektrostroitelstva i planovaia elektrifikatsiia," *Elektrifikatsiia,* 1926, nos. 5–6: 2; "Voprosy elektrostroitelstva i elektrifikatsii," *Elektrifikatsiia,* 1924, nos. 9–10: 19–20; 1925, no. 2: 24; "Uviazka organizatsionnykh i tekhnicheskikh voprosov elektrifikatsii," *Elektrifikatsiia,* 1924, no. 11: 29–30; Elektrobank, *Nekotorye itogi deiatelnosti Elektrobanka,* 16; Ia. Ie. Rubinshtein, *Rol bankov v sovetskom khoziaistve* (Moscow: Gosudarstvennoe izdatelstvo, 1928), 90.

Commissariat of Finance gave Elektrobank the financial carrot to accompany the Glavelektro planning stick; by October 1926, Commissariat of Finance funds constituted 44 percent of Elektrobank's resources and primarily funded regional stations.[152] These formal links greatly strengthened Elektrobank's financial and political base while giving Glavelektro and Gosplan additional indirect control over local utilities. The centralization of financing aided the centralization of control.

With a modus vivendi with the main state banks, major funding from the Commissariat of Finance, and close links with Glavelektro, Elektrobank's only obstacle to total financial control was the Central Communal Bank.[153] Founded in 1924 to provide municipalities with long-term loans, the Central Communal Bank concentrated on housing and city enterprises. In 1924–25, the bank lent 1.2 million rubles for local electrification, one-quarter of its credit for municipal enterprises. In 1925–26, electrification received 5.2 million rubles, a fourfold increase. Although lending an order of magnitude less than Elektrobank, the Central Communal Bank represented a financial base outside Elektrobank's control. If only loans to city utilities are considered, the difference between the banks shrank to a factor of two (Elektrobank's 12.2 million rubles to the Central Communal Bank's 6.4 million rubles), and the newcomer's investment was increasing.[154]

Glavelektro and Elektrobank tried to remove the Central Communal Bank from electrification by charging that it wasted resources with parallel development and that investment in regional stations was more profitable. The Central Communal Bank countered by claiming utilities as an integral part of city economies.[155] Because of its limited assets, the Central Communal Bank could not meet municipal needs, thus leaving urban utilities financially dependent on Elektrobank and Glavelektro.

Elektrobank viewed electrification—and its role—in visionary concepts easily taken from GOELRO and Glavelektro. Initially, it viewed its priorities as agricultural cooperatives, city governments, industry, and regional stations. It quickly reversed these priorities to concentrate on the more productive "factories and workers"; hence, "rural

---

[152] Elektrobank, *Finansirovanie elektrokhoziaistva*, 36, 40.

[153] Elektrobank, *Nekotorye itogi deiatelnosti Elektrobanka*, 16.

[154] "Mestnye finansy," *Kommunalnoe delo*, 1925, no. 1: 21; 1926, no. 7: 43, no. 8: 42–43; Elektrobank, *Finansirovanie elektrokhoziaistva*, 16, 40.

[155] Elektrobank, *Finansirovanie elektrokhoziaistva*, 19; B. Zaitsev, "O kreditovanii kommunalnykh elektrostantsii," *Kommunalnoe delo*, 1926, nos. 21–22: 29–31.

electrification has to take last place."[156] By 1927, Elektrobank had fully adopted the centralized, rationalizing attitude of Glavelektro and sought the "strict coordination" and subordination of all electric stations to regional stations under the rubric of economies of scale.[157]

With major funds coming from the Commissariat of Finance and lesser sums from Gosbank, Glavelektro, and state industries, available credit grew massively from 4.2 million rubles in November 1924 to 74 million rubles in October 1926 (see Table 7.5).[158] Although not offering the ten–twelve years desired by recipients, Elektrobank shifted from short-term to medium-term loans of four or five years, a significant increase from Elektrokredit's three-year maximum. More important, Elektrobank shifted funding priorities from rural and city stations to industrial electrification and the Shatura, Shterov, Cheliabinsk, and Zemo-Avchalsk regional stations.[159] Urban, rural, and kustar stations lost ground to the big producers and consumers of electrical energy. On average, a industrial loan was five times larger than a municipal loan and twelve times larger than a cooperative loan. Despite this bias, Elektrobank loans enabled more than sixty cities and towns to repair, expand, and in some cases construct new stations.[160]

Regional joint stock companies offered another source of combined financial and technical assistance. Their stockholders were Glavelektro, Elektrobank, electrotechnical industries, local governments, cooperatives, and individuals. In the Ukraine, Elektrika had a capital fund of 600,000 rubles, half as much again as the Leningrad-based Elektropomoshch.[161] Small compared with Elektrobank resources, these companies were substantial at the local level.

The state budget remained a minor player in the debates over financing. State funds went directly to regional stations and to such agencies as Glavelektro and the GUKKh. Electrification's share of the state budget hovered steadily around 1.8 percent from 1922 to 1927,

---

[156] Elektrobank, *Finansovoi plan i smeta dokhodov i raskhodov na 1924/25 god* (Moscow: Elektrobank, 1925), 8–9.

[157] Elektrobank, *Finansirovanie elektrokhoziaistva*, 15, and *Nekotorye itogi deiatelnosti Elektrobanka*, 3–5.

[158] Elektrobank, *Nekotorye itogi deiatelnosti Elektrobanka*, 6, and *Finansirovanie elektrokhoziaistva*, 25, 36. Cooperatives invested 250,000 rubles and individuals 74,000 rubles.

[159] "Mestnye finansy," *Kommunalnoe delo*, 1926, no. 4: 60; Elektrobank, *Nekotorye itogi deiatelnosti Elektrobanka*, 11, 14, and *Finansirovanie elektrokhoziaistva*, 40.

[160] Elektrobank, *Nekotorye itogi deiatelnosti Elektrobanka*, 14, 41.

[161] "Voprosy elektrostroitelstva i elektrifikatsii," *Elektrifikatsiia*, 1924, nos. 9–10: 19.

Table 7.5. Elektrobank long-term credit, 1924–26

Recipient	1924–25 (million rubles)[a]	1925–26 (million rubles)[a]
Regional stations	0.02 (0.7)	4.1 (31)
City stations	1.7 (47)	3.8 (29)
Industry	.7 (20)	3.4 (26)
Rural and kustar	1.2 (33)	1.8 (14)
	3.6	13.1

Source: Elektrobank, *Finansirovanie elektrokhoziaistva (dva goda raboty Elektrobanka)* (Moscow-Leningrad, 1927), 40.
[a] Percentage of total in parentheses.

growing from 23 to 102 million rubles.[162] By October 1926, six years after GOELRO began, at least 450 million rubles had been invested in electrification, with 6 percent (29 million rubles) for rural and urban utilities, 29 percent (131 million rubles) for industrial electrification, and 51 percent (229 million rubles) for regional stations (see Table 7.6).[163] This sum accurately reflects state priorities in regional stations and the conversion of industry from steam to electric power. These priorities benefited the first tier at the expense of less industrialized areas.

Despite the increasing funding, demand chronically outstripped supply by at least a factor of ten. State credit for municipal utilities in 1923–24 totaled only 1.4 million rubles for forty-five loans, one-eleventh of the 16 million rubles requested. In 1924–25, the Central Communal Bank received requests for 43 million rubles from municipal enterprises but could fund only 4.7 million. For 1925–26, only seventy-five (10 percent) of more than seven hundred proposals from village and cooperative companies for electrification shared 6 million rubles. For 1926–27, Glavelektro funded only one-quarter of the 46 million rubles in requests for rural electrification.[164]

Inadequate financing did not limit only credit agencies. Glavelektro consistently received only a quarter to a half of its requested budget

[162] Davies, *Soviet Budgetary System*, 83.
[163] Elektrobank, *Finansirovanie elektrokhoziaistva*, 10–12, 40.
[164] "Khronika Tsentra," *Kommunalnoe delo*, 1924, no. 10: 42–43; "Mestnye finansy," *Kommunalnoe delo*, 1926, no. 7: 43; "Khronika kommunalnoi zhizni," *Kommunalnoe delo*, 1926, no. 1: 69; "Mestnye finansy," *Kommunalnoe delo*, 1926, nos. 23–24: 99–100.

Table 7.6. Investment in electrification, 1920–26

Recipient	Million rubles[a]
Regional stations[b]	229 (51)
Electrification of industry	131 (29)
Bank credits[c]	48 (11)
Rural and urban stations[b]	29 (6)
Electrotechnical industry	20 (4)
Railroad electrification	5 (1)
	452

Source: Elektrobank, *Finansirovanie elektrokhoziaistva (dva goda raboty Elektrobanka)* (Moscow-Leningrad, 1927), 10–12, 40.

[a] Percentage of total in parentheses.
[b] From the state budget and banks.
[c] Unidentified by destination, but probably industry.

from the financially strapped government.[165] Inadequate financing adversely affected the pace of electrification by delaying reconstruction, slowing expansion, and hindering new construction, while the centralized control of Elektrobank in cooperation with Glavelektro and Gosplan served as a powerful tool to direct resources toward regional stations and away from the smaller local stations. From the center's point of view, such a move made excellent sense. From the local viewpoint, Elektrobank was strangling the reconstruction and expansion of utilities.

Fuel Choices

DESPITE OFFICIAL ENDORSEMENT of hydropower and peat, utilities returned to coal and oil. In principle, local fuels offered independence from railroads and fuel shortages. In reality, utilities preferred Donets coal and Baku oil, which were easier to handle and burn. As the economy and railroads recovered, so did interest and access to these prewar high-quality fuels. Despite the rhetoric and rationale for local fuels, utilities overwhelmingly continued to use oil and coal.

Local fuels suffered from unfamiliarity, variable quality, inadequate specialized equipment and trained personnel, unattractive costs, and

[165] V. V. Kuibyshev, "Sostoianie elektrotekhnicheskoi promyshlennosti," 44; "Khronika Tsentra," *Kommunalnoe delo*, 1924, nos. 7–8: 38; A. Barilovich, "Plan elektrifikatsii," *Elektrifikatsiia*, 1925, nos. 11–12: 4; I. A. Gladkov, ed., *Perekhod k NEPu: Vosstanovlenie narodnogo khoziaistva SSSR 1921–1925 gg.* (Moscow: Nauka, 1976), 279.

no established organizational framework. Glavelektro could have resolved these deficiencies by a firm institutional commitment to educate utilities, provide skilled staff, improve the quality and availability of equipment, and structure prices. Without such a political commitment, more economic factors prevailed. Indeed, the state actually worked against itself: financial assistance for oil and coal in 1925 exceeded aid to peat and local coal by a factor of six.[166] As a result, prices among fuels differed less in 1924 than in 1913, lessening the attraction of low-quality fuels.[167] Consequently, as utilities recovered, they reduced their fuels from as many as four to one high-quality fuel. In October 1924, conventional fuels powered most utilities: oil and coal fired sixty-three (59 percent) of 107 utilities, wood fired seven stations, and peat one station.[168] Water powered seven stations, with an average size of 500 kW, one-third the thermal station size. As reminders that the civil war "fuel hunger" still lingered, the remaining twenty-nine stations burned both high-quality and low-quality fuels (e.g., oil and wood) with steam turbines and internal combustion engines. Two years later, the number of mixed stations had dropped as fuel supplies improved. A 1925–26 survey of seventy-eight stations found that oil and coal powered 70 percent of capacity directly and 22 percent to some degree.[169] Wood fueled five cities and only eight cities used two or more fuels.

Different institutional viewpoints produced different views of oil. Gosplan and Glavelektro saw the liquid foremost as a "valuta fuel" for export and urged the conversion of oil-fired stations to local fuels.[170] Utilities viewed oil as an excellent fuel because it had the highest energy content of all fuels and was easy to transport, store, and burn. Consequently, oil-fired Diesel engines remained popular among the local and urban stations.[171]

Rural electrifiers envisioned a major role for "green coal" (mini hy-

[166] Fifty-seven million versus ten million rubles; Gorev, "Elektrifikatsiia SSSR," 180.

[167] Electricity from coal cost Elektrotok less than from peat at the new Red October station; ibid., 181. A later report claimed, however, that 1925–26 fuel prices in the Central Industrial Region maintained approximately the same ratio between high- and low-quality fuels as 1913 prices; see M. K. Polivanov, *Raionnye elektricheskie stantsii i podstantsii* (Moscow: Elektro-promyshlennyi kruzhok I. N. Kh. imeni G. V. Plekhanova, 1927), 19.

[168] V. Levi, "Pokazateli raboty elektricheskikh stantsii," *Kommunalnoe delo*, 1926, no. 6: 19, 21–22. Excluded were Moscow, Leningrad, and Baku.

[169] "Biulleten," *Elektrichestvo*, 1927, no. 4: 148.

[170] Levi, "Pervoe Vserossiiskoe soveshchanie," 44.

[171] E. g., Samarkhand, "Kommunalnye predpriiatiia," *Kommunalnoe delo*, 1927, no. 4: 83.

dropower), as opposed to the large-scale "white coal."[172] Green coal proved the major exception to the reversion to coal and oil. Many villages introduced electric light and power by converting a mill or building a small hydroelectric station. Of the Moscow region's thirty rural stations in January 1923, oil fueled thirteen and minihydropower served the remaining seventeen. Since it used existing equipment and peasant services, hydropower was significantly cheaper than oil, providing electricity for 10 kopecks versus 35–50 kopecks per kWh.[173] Minihydropower stations averaged an order of magnitude smaller than thermal stations.[174]

It is ironic but not surprising that only the largest and smallest electric stations in the Soviet Union fully embraced the autarkic concept of local fuels while most utilities remained wedded to their prewar fuels. The color coals—white, gray, brown, and green—powered the newest stations. They could embody the new principles because they began de novo instead of having to modify existing institutions and equipment. The small scale of investment, easy access to supplies, and lack of alternatives made local fuels attractive for rural stations. The investment and push for locally fueled regional stations came from the central government. In the middle stood existing utilities, their equipment and personnel dedicated to the traditional high-quality oil and coal.

Rural Electrification

ALTHOUGH REGIONAL and rural stations shared the common denominator of dependence on the Soviet regime, rural electrification developed as a world apart from the Kashiras and Volkhovs in scale, setting, customers, and economics. Small stations promised political, economic, and social benefits. Politically, rural stations served as visible signs of progress, showing the peasant that the interests of the people were the interests of the party while promoting cooperatives over individual farming.[175] Economically, rural electrification would

[172] V. Levi, "Elektrosnabzhenie Rossii," TsGANKh f. 5208, op. 1, ed. kh. 69, 33.

[173] Ie. Shnirlin, "Elektrifikatsiia Moskovskoi gubernii," Kommunalnoe khoziaistvo, 1924, no. 2: 15.

[174] In Tula gubernia, the hydrostations averaged 9 kW and the thermal stations 100 kW; "Khronika mest," Kommunalnoe delo, 1924, nos. 11–12, 90.

[175] "Elektromotor v derevne," Izvestiia, 9 April 1926, 4.

improve the efficiency of agriculture and revitalize local industry.[176] Socially, electrification would serve as a civilizing agent to close the town–country gap. Despite these proclaimed benefits, the small amount of resources dedicated to rural electrification combined with the problems of literally blazing new trails resulted in only modest gains that fell far short of the visions.

The development of rural electrification was a story of the evolution of local institutions operating within a broader, national framework of legal authority, financial support, industrial suppliers, and propaganda. It was also a tale of local initiative and institutional inadequacies, of Moscow-based organizations venturing forth with detailed plans and returning chastened by the realization that planned development could not effectively be imposed on the countryside. Lacking tsarist precedents and specific knowledge about Western activities, Soviet rural electrification after much trial and error developed guiding principles based on bottom-up initiative to mobilize local resources, supported by higher-level organizations with financial, technical, administrative, and educational assistance. By 1925, administrative and legal procedures were well developed. New stations electrified hundreds of villages, significant progress but only a small fraction of the number needed. Rarely uneventful, rural electrification nonetheless attained an aura of normalcy, complete with standard problems.

The institutions ranged from Glavelektro and Elektrokredit in Moscow to scores of governmental bodies and joint stock companies at the lowest levels supported by regional administrative organizations. The Soviet government did not establish a dedicated high-level body for rural electrification, an indicator of the technology's low priority. The most interested central organization, Glavelektro, treated rural electrification as a secondary matter. Lower-level organizations, such as Elektroselstroi, developed to transfer resources from the center. Cooperatives, a long-standing rural institution trusted by the peasants and ideologically acceptable to the party, provided outside experts and organized joint stock companies that actually operated village utilities. Notably absent from electrification were the Central Bank for Agricultural Credit, which favored other areas of investment such as fertilizers, and the Commissariat of Agriculture, which let its offspring Elektroselstroi carry the torch.[177]

[176] I. Ia. Perelman, *Elektrifikatsiia kustarnoi promyshlennosti* (Moscow: Gosudarstvennoe izdatelstvo, 1921).
[177] I. A. Kirillov, *Tsentralnyi bank selsko-khoziaistvennogo kredita* (Moscow: Kooperativnoe izdatelstvo, 1925), 58–59.

Compared with the GUKKh–Glavelektro battles, rural–center relations were harmonious. In such a youthful field, debates focused on how best to create new organizations rather than on control of utilities. The rural struggles were not as fierce because the actors shared common goals and had less to lose. Rural electrifiers needed all the help they could get, yet their bailiwick was too small to covet. As a "station of local significance" (500 kW or less), the rural station rarely attracted the direct attention of Glavelektro or Gosplan. The lack of a central state body gave local organs more leeway but also limited the flow of resources from Moscow.[178] Instead, a series of decrees and resolutions provided a national framework for small stations.[179] Authority for utility construction and operations devolved to the regional, local, and village levels.[180]

In mid-1924, the SNK approved a model agreement for a joint stock company or a limited-liability partnership to supply electricity.[181] *Elektrifikatsiia* printed model forms so interested groups could begin on a firm legal basis; aspiring electrifiers had only to fill in the blanks and send the form through a specified review process.[182] The application was automatically approved unless an objection was filed within a month.[183] The applying company obtained a monopoly on production and distribution unless superseded by a regional station. In thirty or fifty years, ownership transferred to the government.[184] Glavelektro and the regional ispolkom had rights of technical and economic inspection. These agreements gave the government greater rights than prerevolutionary concessionary agreements, but otherwise they were similar.

Glavelektro feared the construction of stations in disregard of the

---

[178] A. N. Rumiantsev, "Organizatsiia finansirovaniia melkogo elektrostroitelstva i planovaia elektrifikatsiia," *Elektrifikatsiia*, 1926, nos. 5–6: 5.

[179] For an overview, see V. Iurchenko, "Obzor sovetskogo zakonodatelstva po elektrifikatsii," *Elektrifikatsiia*, 1924, nos. 7–8: 3–8, and his "Dekrety i polozhenii po elektrifikatsii," in V. Z. Esin, *Elektrifitsiruite derevniu!* (Moscow: Elektrokredit, 1924), 15–16.

[180] E. g., for Moscow gubernia, see "Po provintsii," *Kommunalnoe khoziaistvo*, 1924, no. 10: 31.

[181] "Osnovnye polozheniia po sostavleniiu dogovorov dlia postroiki i ekspluatatsii elektricheskikh stantsii mestnogo znacheniia tovarishchestvam po elektrosnabzheniiu s ogranichennoi otvetstvennostiu," *Sbornik dekretov, postanovlenii, rasporiazhenii i prikazov po narodnomu khoziaistvu*, 1924, no. 15: 111–14.

[182] See *Elektrifikatsiia*, 1924, nos. 7–8: 8–12; no. 11: 33–37.

[183] "Instruktsiia o poriadke registratsii tovarishchestv po elektrosnabzheniiu s ogranichennoi otvetstvennostiu," *Sbornik dekretov, postanovlenii, rasporiazhenii i prikazov po narodnomu khoziaistvu*, 1925, no. 16: 121–22.

[184] Thirty years for a thermal station and fifty for a hydrostation, reflecting the latter's greater investment; "Khronika Tsentra," *Kommunalnoe delo*, 1924, no. 9: 44.

national plan of electrification. Opening a regional station would waste the investment in a local station. Planning organs did veto stations in these cases.[185] With an eye to the long term, Glavelektro wanted technically compatible stations built to link with future regional networks. Peasants, however, looked at the hard, immediate cost rather than a vague, distant benefit. Instead of paying higher initial costs of AC for an easier transition to a future grid, villages preferred DC.[186] Of 651 rural stations in 1926, only sixty-nine (11 percent) operated on AC.[187] For the small loads and short transmission distances of village electrification, DC stations remained economically more rational than AC. Initial low cost meant near-term application, and a village could switch to AC when the population grew richer.[188]

Funding also proved a bottleneck. The costs of electrifying varied greatly: converting a mill demanded only a few thousand rubles, whereas building a new station and equipping local industries with motors could cost a hundred thousand rubles.[189] Local companies expected and received donations from peasants of labor, materials, and money.[190] Nonetheless, rural electrification needed outside financing.[191] Rarely was only money provided. Creditors quickly discovered the necessity of packaging technical and managerial support with financing, a lesson learned decades earlier in the West.

Rural electrifiers expected the central government to provide organizational assistance, financing, material aid, and propaganda.[192]

---

[185] One of Elektroekspluatatsiia's initial agreements in 1923 was voided because the Kashira regional station would soon supply the region; Elektroekspluatatsiia, *Otchet pravleniia aktsionernogo obshchestva elektricheskikh predpriiatii 'elektroekspluatatsiia' obshchemu sobraniiu aktsionerov 29 dekabria 1924 g. za pervyi operatsionnyi god (5 iiunia 1923 g. - 1 oktiabria 1924 g.)* (Moscow: Elektroekspluatatsiia, 1925), 12.

[186] P. Skvortsov, "Novye zadachi elektrosnabzheniia," *Kommunalnoe delo*, 1924, no. 6: 20–23; F. Dits, "Elektrifikatsiia derevni," *Kommunalnoe delo*, 1926, no. 2: 21.

[187] "Biulleten," *Elektrifikatsiia*, 1928, no. 11: 38. The current of ninety-nine stations was unknown.

[188] Iu. V. Skobeltsyn, "Snabzhenie elektricheskoi energiei selskikh mestnostei ot stantsii maloi moshchnosti," *Elektrifikatsiia*, 1924, no. 4: 211; Aktsionernoe obshchestvo po elektrifikatsii selskogo khoziaistva, *Elektroselstroi i ego deiatelnost* (Moscow: Novaia derevnia, 1924), 15–16.

[189] A. N. Rumiantsev, "Stolko stoit elektrifitsirovat derevniu," in Elektrokredit, *Iz praktiki kooperativnoi elektrifikatsii*, 8; V. I. Moshkevich, "Stoimost elektrifikatsii derevni," in Esin, *Elektrifitsiruite derevniu!*, 7.

[190] *Elektroselstroi i ego deiatelnost*, 6.

[191] *Trudy 1-go Vsesoiuznogo sezda po selsko-khoziaistvennomu kreditu 15–21 dekabria 1924* (Moscow: Izdanie Tsentralnogo S.-Kh. Banka Soiuza S.S.S.R., 1925), 329–30, 407–8.

[192] A. Kulikovskii, "O role gosudarstva v elektrifikatsii derevni," *Ekonomicheskaia zhizn*, 22 April 1922, 1.

Glavelektro and Elektrokredit promoted rural electrification. In 1923, they published *Elektrifikatsiia*, a propagandizing, popular science journal for rural electrification. In addition to funding, Elektrokredit offered technical and administrative assistance and encouraging advice, such as including a wide range of politically important and "literate people," such as teachers, in preparatory work to create a network of powerful supporters. Elektrokredit warned, "The path to electrification will have many obstacles and the main one is lack of money, but do not lose heart. What can't be done in one year perhaps can be done in two or three."[193]

To assist rural electrification, industry and government supported regional companies. These firms—underfunded, inexperienced, and caught in changing political currents—provided services to customers even less knowledgeable. The most visible firms were Elektroselstroi (Electro-agricultural Construction) and Elektroekspluatatsiia (Electric Operation), both based in Moscow, Elektrika (Electric) in the Ukraine, Elektropomoshch (Electro-help) in the north, and Elektrokrai (Electro-region) in the north Caucasus. By 1925, over one hundred such firms organized local companies, provided technical inspections and advice, aided financing, ordered equipment and materials, and assisted construction.[194] An independent Moscow-based group, the Society to Assist Electrification (Nauchno-Tekhnicheskoe Obshchestvo Sodeistviia Elektrifikatsii), tried to function as a mini-TsES. According to a commentary on the problems facing rural electrifiers, the society suffered from overworked members, poor links with provincial workers, extremely limited resources, and a one-year delay in having its statutes approved.[195]

Cooperatives and local joint stock companies actually brought the electric station to a small town or rural region. Their shares were subscribed by the local ispolkom, cooperatives, individuals, and regional bodies. Peasants could buy inexpensive shares, which provided additional capital and created a base of committed individuals.[196] Additional funding came from the better-financed regional companies, themselves financed by industrial and state funds.

[193] E. g., Instruktor, "Kak organizirovat tovarishchestvo po elektrifikatsii," *Iz praktiki kooperativnoi elektrifikatsii*, 7.
[194] *Elektroselstroi i ego deiatelnost*, 8; *Perekhod k NEPu*, 289.
[195] A. P. Kravchenko, "Nauchno-tekhnicheskoe obshchestvo sodeistviia elektrifikatsii," *Elektrifikatsiia*, 1925, nos. 11–12, 34.
[196] The Ukrainian Vintzelelektro sold shares for 100 rubles to the regional ispolkom and 10 rubles to peasants; "Korrespondentsii," *Kommunalnoe delo*, 1925, no. 1: 49.

Cooperatives played a major role in introducing electricity.[197] For example, Tsentrosoiuzkartofel (the central union of potato cooperatives), funded by Elektrokredit, built seventeen stations in the Central Industrial Region from 1920 to 1924 for processing potatoes and lighting. Kustar handicrafts provided the low-technology materials and equipment needed for these small utilities. The value of kustar electrotechnical goods from 1921–22 to 1925–26 increased sixfold to 24 million rubles, or one-fifth of all electrotechnical goods.[198] Local companies handled the actual construction, and villagers contributed labor, materials such as wooden poles, and sometimes money. By December 1926, cooperatives owned and operated nearly half of the 651 rural stations.[199]

The activities of Elektroselstroi and Elektroekspluatatsiia provide a telling glimpse into the demands of small-scale electrification. Both endured harsh teething troubles. One remained committed to rural electrification, but the other became a convert to regional stations and existing utilities. As part of its restructuring under the NEP, the Commissariat of Agriculture transformed its section for rural electrification, Elektrozem, into Elektroselstroi in June 1922. Elektrozem's failure to develop the rural equivalent of the tsarist model agreement may have contributed to this change.[200] Elektroselstroi's first year was dominated by efforts to obtain its promised financial and material dowry and to repulse an amalgamation attempt by Gosselsindikat, a newly created joint stock company also spawned from the Commissariat of Agriculture. The main operations were in the Moscow region, though Elektroselstroi had a Petrograd office and three colonies for electric plow research.[201]

In late 1923, Elektroselstroi started negotiations on cooperation with ASEA, the Swedish General Electric Company.[202] In June 1924, the

[197] A. Kravchenko, "Ot slov k delu," and V. Pleskov, "Kooperativnaia i mestnaia elektrifikatsiia," in *Iz praktiki kooperativnoi elektrifikatsii*, 3, 4.

[198] Elektrobank, *Finansirovanie elektrokhoziaistva*, 6.

[199] "Khronika mest," *Kommunalnoe delo*, 1924, no. 5: 53; "Elektrifikatsiia v kartofelnykh raionakh," *Iz praktiki kooperativnoi elektrifikatsii*, 9; M. Tipograf, "Elektrotekhnicheskaia promyshlennost," in A. M. Ginzberg, ed., *Chastnyi kapital v narodnom khoziaistve SSSR* (Moscow: Promizdat, 1927), 548; A. P. Kravchenko, "Rabota Selskosoiuza v oblasti elektrifikatsii," *Elektrifikatsiia*, 1927, no. 2: 20.

[200] A. Kulikovskii, "O role gosudarstva," 1.

[201] Elektroselstroi, *Gosudarstvennaia montazhno-stroitelnaia kontora po elektrifikatsii selskogo khoziaistva 'Elektroselstroi' za 1 1/2 goda ee sushchestvovaniia* (Moscow: Narkom zemledeliia, 1923), 4–6, 8, 25–31. See also *Elektroselstroi i ego deiatelnost*, 30.

[202] This is probably the General Electric cooperation listed by Antony C. Sutton in *Western Technology and Soviet Economic Development* (Stanford: Hoover Institution Publications, 1968), 186.

government firm became a 2 million ruble joint stock company with access to foreign funding, technology, and experience. Elektroselstroi worked closely with Elektrokredit, the Commissariat of Agriculture, and private and German sources of capital and technology.[203] Elektroselstroi received 251 inquiries in its first eighteen months, of which 103 turned into clients and 28 signed contracts to draw up projects and build stations.[204] If fully implemented, the projects would cost 2.9 million rubles, equivalent to Elektrokredit's total loans in 1923, but cash flow problems canceled some projects.[205]

Elektroekspluatatsiia was formed by the Moscow Soviet in June 1923 but was sent into the world without theoretical or practical guidelines. Important questions about relations with Glavelektro and the source and nature of credit remained unanswered one year later, but the firm had learned what not to do. Funded primarily by future station owners and electrical equipment manufacturers, it quickly learned about the drawbacks of bartering and the need to obtain loans before the summer construction season, when money arriving a month late could cause a year's delay.[206] Initially, Elektroekspluatatsiia operated like a concession, seeking 30- to 36-year monopoly agreements, but it soon abandoned this approach on grounds of inadequate foreign funding and the inherent dangers of monopolies. More likely, foreign investors were not interested in long-term concessions for small Soviet stations. Elektroekspluatatsiia quickly focused on projects that promised a full return on capital within ten years, twice Elektrobank's timeframe. Clients, not outsiders, would invest in a station constructed and controlled by Elektroekspluatatsiia and operated by the client. Station control would revert to the client when the loan was repaid.[207]

Despite networks of representatives and 612 inquiries, Elektroekspluatatsiia signed only sixteen construction projects and eight hydrological surveys in its first year. The company was the equivalent of a mine canary, finding problems everywhere. Local craft guilds proved costly and produced poor work; state electrotechnical trusts were unreliable; obtaining credit proved difficult; the legal framework

[203] Elektroselstroi, *Gosudarstvennaia montazhno-stroitelnaia kontora*, 18–19, 23; "Mestnaia elektrifikatsiia," *Elektrifikatsiia*, 1924, nos. 9–10: 37; *Elektroselstroi i ego deiatelnost*, 7–8.

[204] Seventeen agreements to build stations, six for planning work, four to operate electric plows, and one to build and operate a concessionary station.

[205] Elektroselstroi, *Gosudarstvennaia montazhno-stroitelnaia kontora*, 15, 21, 23.

[206] Elektroekspluatatsiia, *Otchet pravleniia*, 4, 13–14, 16–18, 29.

[207] Ibid., 5–6.

remained uncertain; relations with its main clients, the ispolkom, were fraught with conflict, particularly over money; relations with Glavelektro and its organs proved equally demanding; the stations did not operate as efficiently as predicted; and qualified personnel were extremely rare. This Job-like litany of woes offers an idea of the problems inherent in local electrification. It comes as no surprise that by 1925 Elektroekspluatatsiia supported regional stations wholeheartedly over local stations and refocused its activities on areas with existing stations and a developed kustar industry.[208]

Elektroselstroi and Elektroekspluatatsiia discovered that local electrification could not be dictated from above—that it grew best from below, aided by outside organizational, technical, and financial assistance. Local initiative existed in the countryside, but the technical and economic means and social organization did not. Introducing technical change successfully required institutions to transfer the technology and arrange for its reception. Affordable credit and equipment were more important than comprehensive plans.[209] These firms learned the hard way that the economic feasibility of stations depended on industrial consumption as well as lighting. But there would be no industry without inexpensive electrical equipment and inexpensive railroad transportation to foster trade with the city.[210] Rural electrification found itself in the same fix as regional stations: increasing productivity was impossible without the assistance of other sectors of the economy. The cart could not come before the horse.

Despite these problems, electric light and power did penetrate the countryside. Success stories filled *Elektrifikatsiia*—a rural handicrafts shop equipped with electric motors, peasant homes illuminated by electric light. After five years of GOELRO, electricity flowed in 1,150 villages and 84,000 homes.[211] Although nearly a fourfold increase since 1917, this still covered only a small fraction of Russia's 84 million peasants in 18 million households.[212] Together, the 651 independent rural stations, 77 factory stations, and 140 substations connected to a regional station supplied 17.9 MW in 1926, a tenfold increase in less than a decade but still "only a drop in the sea. . . . we need tens of tens of thousands of electric stations with hundreds of thousands

[208] Ibid., 7, 9–11, 20–21, 22–26.
[209] A. Smirnov, "Zadachi selskoi elektrifikatsii," in *Elektroselstroi i ego deiatelnost*, 29–30.
[210] Esin, *Elektrifitsiruite derevniu!*, 3–5; *Elektroselstroi i ego deiatelnost*, 4.
[211] A. A. Kulikovskii, "Nekotorye tsifry, kharakteriziruiushchie selskoe elektrosnabzhenie SSSR," *Elektrifikatsiia*, 1926, no. 12: 20.
[212] "Voprosy elektrostroitelstva," *Elektrifikatsiia*, 1924, nos. 9–10: 19.

of electric motors."[213] The electrotechnical chasm between town and country had only started to be bridged.

By 1926, experience and the NEP's market orientation had significantly revised electrification's intended rural role. The political and social goals of transforming the countryside had slipped behind the economic imperatives of increasing station efficiency and profitability. The basic problems of poverty, a technologically illiterate peasantry, and inadequate credit remained, but increasing load factors had become equally important. Electric lighting alone did not a profitable (or at least loan-repaying) enterprise make—that required daytime industrial loads.[214] The economics of efficiency had filtered down to the countryside, pushed by the demands of making a return on investment.

Emphasizing broad structural policies remarkably similar to those of the American reindustrialization debate of the 1980s, Glavelektro urged a state focus on tax policy, transportation, and other indirect measures to assist the economic development of cooperatives.[215] This reorientation reflected a realization by some rurally oriented electrifiers that detailed plans were pointless and that the state should improve the country's infrastructure instead of dictating plans.[216] Economic criteria had undermined social goals in the absence of a strong political commitment.

The Foreign Role

THE FLOW OF Western electrotechnology resumed during the NEP, but neither on the scale desired nor with the prewar degree of foreign control. Foreign firms now dealt with the Commissariat of Foreign Trade as well as with state trusts and firms. Independent foreign investment ceased to exist; Western firms entered into state-approved agreements. Soviets, not foreigners, made the decisions. The advisers and critical equipment, however, remained Western.

GOELRO assumed, with important caveats, the participation of foreign financing and technology. With the exception of ASEA's agree-

---

[213] "Biulleten," *Elektrifikatsiia*, 1928, no. 11: 39; E. N. Moiseenko-Velikaia-Gorev, "Proizvodstvo elektricheskoi energii," *Elektrifikatsiia*, 1923, no. 8: 7.

[214] F. Dits, "Elektrifikatsiia derevni," *Kommunalnoe delo*, 1926, no. 2: 20–21.

[215] E. g., Lester Thurow, *The Zero-Sum Solution: Building a World-Class American Economy* (New York: Simon and Schuster, 1985).

[216] A. Z. Goltsman and A. A. Gorev, "Plan elektrifikatsii i krestianskoe khoziaistvo," *Planovoe khoziaistvo*, 1925, no. 4: 182–83.

Table 7.7. Electrotechnical sales, 1913–26

Year	Domestic[a]		Imports[a]	Total
	State	Local[b]		
1913	80 (60)		54 (40)	134
1916	96 (71)		39 (29)	135
1921–22	5 (24)	4 (19)	12 (57)	21
1922–23	16 (44)	6 (17)	14 (39)	36
1923–24	40 (63)	9 (14)	15 (23)	64
1924–25	54 (61)	14 (16)	21 (24)	89
1925–26	85 (62)	24 (18)	28 (20)	137
Postwar total	200 (58)	57 (16)	90 (26)	347

Source: Elektrobank, *Finansirovanie elektrokhoziaistva (dva goda raboty Elektrobanka)* (Moscow-Leningrad, 1927), 6.
[a] Percentage of all sales in parentheses.
[b] Kustar industry and other nonstate enterprises.

ment with Elektroselstroi, major foreign involvement did not occur. The political difficulties of diplomatic recognition, Soviet repudiation of tsarist foreign debt, nationalization of foreign industries, unattractive concessions, and fear of Bolshevik-fermented revolution made foreigners reluctant to invest and foreign governments hesitant to offer trade privileges.[217] Soviet financial policy, which tried to maximize exports and minimize imports, further restricted trade, as did Communist fear of an alliance between foreign industrialists and Russian peasants whereby inexpensive foreign goods would undercut state industries.[218]

Nonetheless, foreign firms sent technologists and technologies to Russia. The pages of *Elektrichestvo* again filled with advertisements from familiar firms: Siemens and Halske, AEG, Metropolitan-Vickers, General Electric, and ASEA. As Table 7.7 shows, imports more than doubled in value from 1921–22 to 1925–26, but their market share dropped sharply from 57 to 20 percent as domestic production recovered and then surpassed prerevolutionary output.[219] Although imports had a smaller market share and value than in 1913, they played

[217] Carr, *Bolshevik Revolution*, 1: 276–89.
[218] Charles Bettelheim, *Class Struggles in the USSR: The Second Period, 1923–1930* (New York: Monthly Review Press, 1978), 58–59; Goltsman and Gorev, "Plan elektrifikatsii," 178.
[219] Elektrobank, *Finansirovanie elektrokhoziaistva*, 6.

Table 7.8. Imports for regional stations

Station	Equipment
Kashira	two 6-MW Brown-Boveri turbogenerators Babcox and Wilcox, Garbe, Sterling boilers
Nizhegorod	six Babcox and Wilcox boilers two 10-MW AEG turbogenerators
Red October	10-MW prewar Brown-Boveri turbogenerator Walter boilers with Makarev fireboxes Bruno boilers 10-MW Czech turbogenerator
Shatura	three 16-MW Bruno turbogenerators Siemens generators Garbe boilers
Shterov	two 10-MW Vickers turbogenerators six Babcox and Wilcox boilers Combustion Rationelle fireboxes
Volkhov	four 12-MW ASEA turbines

Source: Glavelektro, *Obzor sostoianiia rabot po krupnomu elektrostroitelstvu na 1ogo oktiabria 1925 g.* (Moscow, 1926), 19–33.

a vital role in the restoration of utilities and construction of regional stations.

Imports were essential for Soviet electrification. The spare parts that restored utilities in the early 1920s came from abroad. Urban stations sought imports because they were less expensive, more reliable, delivered faster, and built better than their domestic equivalents.[220] More important from Glavelektro's perspective, regional stations depended on foreign technology and expertise. As Table 7.8 illustrates, imports provided key components for six of the first seven regional stations. Only Kizel did not import equipment; instead, it used machinery from the old Oranienbaum station. Without Western technology, the GOELRO plan could not be realized.

The West, defined increasingly as the United States, continued to serve as a model as well as a source of technology. What better affirmation of the vigor of Soviet construction could there be than to label it American? At the Volkhov hydrostation, a propagandist proclaimed, "Here you see the current America—noise, thunder—all, all

---

[220] E. g., Tomsk received a foreign turbogenerator more quickly at half the cost of a Soviet system; "Kommunalnye predpriiatiia," *Kommunalnoe delo*, 1925, no. 20: 53.

America. In a word, there is not a Russian approach but an American tempo."[221] Yet, even as Fordism, Taylorism, and "Amerikanizm" captured the enthusiasm of Soviet modernizers, the decreased reliance on German technology presaged a shift in priorities from the technologies of the second industrial revolution to those of the first—from the science-based chemical and electrical technologies to metallurgy and mining.[222]

The GOELRO Plan, 1924–1926: Repudiation and Reaffirmation

THE 1924–26 period was crucial for all three tiers of utilities as the Soviet government had to answer the fundamental question whether local governments and profitability or the state should determine the pace and path of industrial development.[223] Under what rules would electrification operate? Would municipalities and the market, serving priorities not necessarily the state's, or the party drive the development of electrification? One direction was decentralized electrification based on local demand and regional stations only in a distant future. Rural and urban utilities supported this approach; for them, short-term economics were major considerations. The other direction was centralized electrification, based on regional stations and the GOELRO plan. Gosplan, Glavelektro, Elektrobank, and the Communist party supported this top-down development. Ultimately, this was a political question, decided by the party.

By 1925, a normalcy of sorts had returned as generation surpassed the 1913 level.[224] The immediate post–civil war shortages had eased and administrative structures had stabilized. Most utilities had completed major repairs and reconstruction and turned to expanding their capacity and clientele. Instead of shortages of food and fuel, utilities now suffered from shortages of credit, skilled personnel, and materials, although high fuel prices, uneconomical tariffs, and poor statistical and accounting procedures also restrained expansion. The failure of the domestic electrotechnical industry to fulfill orders had

[221] P. I. Voevodin, "Na Volkhovstroe," *Elektrifikatsiia*, 1924, no. 2: 24.

[222] Hans Rogger, "*Amerikanizm* and the Economic Development of Russia," *Journal for the Comparative Study of Society and History*, 23 (1981): 388–89; Thomas P. Hughes, *American Genesis: A Century of Invention and Technological Enthusiasm* (New York: Viking, 1989), 249–84; Kendall E. Bailes, "The American Connection: Ideology and the Transfer of American Technology to the Soviet Union, 1917–1941," *Journal for the Comparative Study of Society and History*, 23 (1981): 429.

[223] Bettelheim, *Class Struggles*, 278.

[224] "Biulleten," *Elektrichestvo*, 1927, no. 1: 43.

also placed a "series of electric stations in an extremely grave situation and caused them enormous losses." Inferior materials, inadequate testing, and poor construction were not unknown, and utilities learned the hard way to test equipment at the factory before delivery.[225] Although not trivial, these shortages were normal problems of capital and economics with prerevolutionary and Western antecedents.

In attempting to expand, urban utilities began to feel their secondary status compared with regional stations. The lower tiers rebelled against the GOELRO plan in 1924–25 but contained their actions within narrow administrative bounds, ensuring eventual defeat. This repudiation of the centralized electrification of the GOELRO plan was a remarkable outburst of independence by the utilities, which viewed Glavelektro as pursuing a distant dream at their immediate expense.

In June 1924, the All-Union Conference on Electricity Supply met in Moscow, the first major electrotechnical meeting since the 8th All-Russian Electrotechnical Conference in October 1921. Glavelektro and the GUKKh had planned the conference since July 1923 to improve local–center relations, establish standards and procedures, and help rationalize the electric supply.[226] The conference certainly improved local–center communications, if not relations. Glavelektro and the local utilities discovered that they operated on different wavelengths. The center talked about future promises, while the utilities saw those promises contributing to their present problems.

From the local perspective, regional stations consumed too much and returned too little. By contrast, local stations provided a quicker return on investment and served a much wider population. The delegates called for an alternate, decentralized approach to replace the GOELRO plan: "The basic task of electrification in the country now lies in the reconstruction and construction of local stations, meeting existing demands, and, in turn, providing an unfocused market for the fuel and metal industries. By the construction of medium stations according to established technical norms, we create users for future regional stations and thus prepare for large-scale electrification."[227]

The key issue now was not direct control but financing, followed by

[225] "Kommunalnoe predpriiatiia," *Kommunalnoe delo*, 1925, no. 1: 30–31; Levi, "Pervoe Vserossiiskoe soveshchanie," 46–47, 49; "Khronika Tsentra," *Kommunalnoe delo*, 1924, no. 10: 67.
[226] G. O. Levit, "35 let Vsesoiuznoi konferentsii po elektrosnabzheniiu," *Elektrichestvo*, 1959, no. 12: 75; *Biulleten Pervoi Vsesoiuznoi konferentsii po elektrosnabzheniiu pri Glavelektro VSNKh i GUKKh RSFSR* (Moscow: Glavelektro, 1924), 1, 3–4.
[227] Iu. Mitlianskii, "Vsesoiuznaia konferentsiia po elektrosnabzheniiu i mestnye stantsii," *Kommunalnoe delo*, 1924, nos. 7–8: 20.

materials and personnel. The delegates declared the "situation of provincial stations extremely serious, in several cases verging on the catastrophic, owing to a complete lack of credit needed for reconstruction."[228] To solve the funding squeeze, the conference recommended redirecting state funds from regional to local stations. Existing construction, such as the Volkhov hydrostation, should be completed, but new projects, such as the Svir hydrostation, should be halted until new small stations had satisfied local needs.[229] The conference did not reject regional stations but relegated them to a future, more advanced phase of development. Rich countries could afford to construct regional stations first and then bring industry to them, but "we should be careful of such and should build only where the market is ripe."[230] In the Soviet Union, local stations would prepare the way by raising electrical consumption first. Regional stations would substitute local for imported fuels, not electrify unindustrialized areas.

The utilities declared that they needed more money but with less state supervision. Municipal decisions, made without the "guardianship and interference" of central organs, would ensure the best use of funds.[231] More long-term credit, combined with an increase of the capital of the Central Communal Bank and the reconstruction (*perestroika*) of the Russian electrotechnical industry, would advance local electrification far faster than would regional stations. Foreign equipment and capital would further increase the tempo of growth. The utilities also sought more money from local governments, industry, and other present and future users, including prompt payment by the military and local governments, to ease daily operations.[232]

The repudiation of GOELRO extended to its roots in the prewar fuel crises. One speaker advocated building local stations specifically to assist the coal industry.[233] This argument stood firmly against the events and technocratic engineering mindset of the previous decade,

---

[228] Ibid., 19.

[229] "Rezoliutsii Pervoi Vsesoiuznoi konferentsii po elektrosnabzheniiu," *Elektrichestvo*, 1924, no. 9 (conference supplement): 3.

[230] V. Levi, "Mestnoe elektrostroitelstvo i elektrifikatsionnaia politika," *Kommunalnoe delo*, 1925, nos. 13–14: 11. Levi rhetorically asked, what if Leningrad had received 40 million rubles instead of the still incomplete Volkhov? He concluded that electric supply would have increased more quickly but hastened to note that the hydrostation was not a mistake, "as its importance is not only economic."

[231] "Rezoliutsii Pervoi," 15. City utilities would remain in the GUKKh framework, however, leaving no doubt where the interference came from.

[232] Ibid., 1, 15–17; D. Sheinis, "Podgotovka k sezdu," *Kommunalnoe delo*, 1925, no. 1: 3.

[233] Mitlianskii, "Vsesoiuznaia konferentsiia," 20.

though one year later Elektroplan followed by contemplating a large conventionally fueled station for Moscow.

Other conferences in 1924–25 supported the reallocation of resources from regional to urban stations.[234] By their preference for the Central Communal Bank over Elektrobank as the credit center for electrification, the utilities firmly declared that they were an integral part of their local economies, a position supported by municipal governments.[235] The firm bias by utilities against regional stations contrasted sharply to the increasing centralization of authority among Glavelektro, Gosplan, and Elektrobank.

Within Glavelektro, interest had increased in supporting conventional stations, partly because the implementation of the GOELRO plan had proceeded far more slowly than desired. A 1925 overview by Krzhizhanovskii, Shulgin, and V. Z. Esin, three founders who moved to Gosplan, showed that the plan's creators viewed it quite differently than in 1920. A more militant and political yet pragmatic view dominated, reflecting the changed environment and electrification's evolving status. Regional stations remained the key electrification technology, but small rural stations gained in status based on political instead of economic and social goals. Simultaneously, Shulgin stated that electrification's role in industrial development should be determined by immediate demand, return on investment, and other economic factors. Central tasks should be limited to state-scale transportation systems and regional stations.[236] These principles agreed with the utilities' major demands, not with the concepts of a command economy. None of these viewpoints was new; their acceptance by the Gosplan and Glavelektro leadership, however, was.

As Table 7.9 shows, GOELRO achieved 10 percent of its planned capacity in its first six years after the completion of the Volkhov. Although, the 164 MW was less than half of the 314 MW originally planned for those seven stations (see Table 7.1), it was twice as great as the 80 MW added to urban and rural stations since 1913.[237] The

---

[234] E. g., the first congress on the Tadzhikistan electric economy ("Khronika Tsentra," *Kommunalnoe delo*, 1924, nos. 11–12: 75) and the First All-Russian Conference of Communal Electric Station Managers (Levi, "Pervoe Vserossiiskoe soveshchanie," 42–43, 48–49; "Ofitsialnyi otdel," *Kommunalnoe delo*, 1926, no. 7: 73).

[235] Levi, "Pervoe Vserossiiskoe soveshchanie," 42, 48–49; "Kommunalnye predpriiatiia," *Kommunalnoe delo*, 1925, no. 1: 34.

[236] E. Ia. Shulgin, "K peresmotru plana elektrifikatsii," *Planovoe khoziaistvo*, 1925, no. 2: 22–23; "Kommunalnye predpriiatiia," *Kommunalnoe delo*, 1925, no. 5: 61.

[237] "Biulleten," *Elektrichestvo*, 1927, no. 1: 43.

Table 7.9. Status of regional stations, October 1925

Status	Stations	Capacity (MW)[a]
Operating, under construction	Kashira, Kizel, Nizhegorod, Red October, Shatura, Shterov, Volkhov	164 (11)
Just starting construction	Cheliabinsk, Saratov, Kharkov	235 (16)
Planning phase	Dniepr, Svir	350 (24)
Not started	15 stations	695 (48)
	27 stations	1,444

Source: Glavelektro, *Obzor sostoianiia rabot po krupnomu elektrostroitelstvu na 1ogo ok-tiabria 1925 g.* (Moscow, 1926), 6, 8.
[a] Percentage of planned capacity in parentheses.

concentration of resources on regional stations had produced corresponding results. Only half of GOELRO's regional stations would be operating in 1930, according to Gorev's prognosis in January 1925.[238]

The 1920 GOELRO plan was polemical, but it was the polemic of technocrats and engineers transforming society. The more traditional political polemics of the 1925 Krzhizhanovskii represented party more than technocratic concepts as he outlined a Soviet victory over capitalism. Electrification remained the answer, but now the question was who would win the worldwide economic war for the future. The West remained both a source of emulation and a threat, with "Ford and his system the main elements of this struggle against us." The "rationalization of the Western economy on the basis of the development of energy has turned a special blade against us," a danger to be deflected by a planned, nationalized Soviet economy reconstructed on a rational energy base.[239]

Five years of a state plan for electrification had not transformed the country but instead revealed the dependence of electrification on other sectors of the economy. Implementation, Krzhizhanovskii acknowledged, had fulfilled less than half its goals because of the poor economy and the opposition caused by the "enormous inertia of all

[238] Gorev, "Elektrifikatsiia SSSR," 178.
[239] G. M. Krzhizhanovskii, "Perspektivy elektrifikatsii," in G. M. Krzhizhanovskii, A. A. Gorev, and V. Z. Esin, *Chetyre goda elektrifikatsii SSSR* (Moscow: Planovoe khoziaistvo, 1925), 15, 21.

those hoary habits of thought which are so natural in [this] agrarian, economically backward country."[240] Poor harvests, slower railroad reconstruction, and less oil production than predicted had cut economic growth and exports.[241] Unmentioned but equally important was state investment in electrification, which reached less than half of what GOELRO had planned. Measured against national goals, progress was poor. Measured against postwar Western advances, Russia had actually fallen farther behind, though Western engineers agreed it was on the right track.[242]

In 1921, the Communist party assumed that the elimination of the town–country gap would promote the transition to communism. As the NEP failed to remove this gap, the state and party leadership realized that the survival of the state depended on integrating the peasants into the national economy.[243] For Krzhizhanovskii, the issue was simple: "Unless we want to be turned into a colony of Western capitalism, we need the decisive and quickest industrialization and reconstruction, but this means mastering the peasant market so that wide layers of peasants can see the advantages of a large-scale state economy."[244] Improving the peasant economy demanded rising above the "three backward forms of agriculture"—the wooden plow, sickle, and flail—by the rapid growth of small stations, whose political significance was now "impossible to underestimate."[245] Political needs demanded rural electrification, despite its financial problems. As with regional stations, political and social goals dictated less rigorous economic criteria.

Despite the slow realization of regional stations and strong pressure from below for local stations, Glavelektro and Gosplan continued to focus on regional stations, the GOELRO raison d'être. The partial retreat sounded by Glavelektro and Gosplan proved temporary as the party debate in 1924–26 about the future economy moved from the NEP toward more directed, large-scale industrialization.[246] The Com-

[240] Ibid., 12, 9.

[241] Shulgin, "K peresmotru," 22–23.

[242] More precisely, "the overall thrust of the London [1924 World Power Conference] is that we were completely right"; Krzhizhanovskii, Gorev, and Esin, *Chetyre goda*, 13–14.

[243] R. W. Davies, *The Socialist Offensive: The Collectivization of Soviet Agriculture, 1929–1930* (Cambridge: Harvard University Press, 1980), 28–38.

[244] Krzhizhanovskii, "Perspektivy elektrifikatsii," in Krzhizhanovskii, Gorev, and Esin, *Chetyre goda*, 15.

[245] Ibid., 16; V. Z. Esin, "Elektrifikatsiia i mekhanizatsiia v selskom khoziaistve," in ibid., 37.

[246] E. H. Carr, *Socialism in One Country, 1924–1926*, vol. 1 (New York: Macmillan,

munist party announced the end of restoration (*vosstanovlenie*) and a
shift to reconstruction (*rekonstruktsiia*) in 1925–26 with greatly increased investment in large-scale industrial enterprises.[247] The 14th
Party Congress in December 1925, "a decisive landmark in the progress of Soviet planning" according to E. H. Carr, approved this rapid,
planned expansion of industry to underpin economically socialism in
one country.[248] This decision renewed promotion of large-scale projects and the control of economic development by credit. In both
areas, Gosplan, Glavelektro, and Elektrobank led the way.

To conform to the changing political and economic environment
and Gosplan's 1926–30 control figures, Glavelektro reviewed the original GOELRO plan in early 1925.[249] Gorev, head of the section on energy, presented this variant to the Gosplan presidium in July 1926,
several months behind schedule. The Gosplan presidium, chaired by
Krzhizhanovskii, approved the proposal as "the first approximation
of a long-term plan of electrification subject to more precise definition
with the completion of the development of a long-term general plan
of the national economy."[250] In December 1926, Krzhizhanovskii and
Gorev presented Gosplan's ambitious five-year electrification plan to
the SNK. The two engineers stated that the new electrification plan
would link closely with other developmental plans not yet elucidated.
The 1926–31 plan called for 930 million rubles to construct regional
stations totaling 1,278 MW, a nearly fivefold increase in funding. The
capacity of local utilities would double from 350 to 754 MW, and that
of industrial stations would increase from 750 to 1,000 MW. This expansion would provide the energy for industry not yet created; social
transformation would be a secondary goal.[251]

1958), 352–53; Cohen, *Bukharin and the Bolshevik Revolution*, 160–212, 243–69; Alexander
Elrich, *The Soviet Industrialization Debate, 1924–1928* (Cambridge: Harvard University
Press, 1960); Moshe Lewin, *Political Undercurrents in Soviet Economic Debates* (Princeton:
Princeton University Press, 1974); Peter Rutland, *The Myth of the Plan: Lessons of Soviet
Planning Experience* (London: Hutchinson, 1985), 68–82; Keith Smith, "Economic Theory
and the Closure of the Soviet Industrialization Debate," in Keith Smith, ed., *Soviet
Industrialization and Soviet Maturity* (London: Routledge and Kegan Paul, 1986), 23–49.
[247] Richard Gregor, ed., *Resolutions and Decisions of the Communist Party of the Soviet
Union: The Early Soviet Period, 1917–1929*, vol. 2 (Toronto: University of Toronto Press,
1974); Carr and Davies, *Foundations of a Planned Economy*, 271–74.
[248] Carr, *Socialism in One Country*, 508; see also R. W. Davies, *The Soviet Economy in
Turmoil, 1929–1930* (Cambridge: Harvard University Press, 1989), 47–49.
[249] "Khronika," *Elektrichestvo*, 1925, no. 2: 126–27.
[250] "Perspektivnyi plan elektrifikatsii raionov," *Ekonomicheskaia zhizn*, 31 July 1926, 1.
[251] Ibid. The article's numbers are internally inconsistent. See also "Perspektivy elektrifikatsii SSSR," *Ekonomicheskaia zhizn*, 12 December 1926, 1.

In a victory for advocates of regional stations and forced industrialization, Glavelektro under Trotsky in July 1925 began preparing a six-
year, 130 million ruble plan for the massive Dniepr hydrostation.[252] In
a complex blend of Ukrainian nationalism, industrialization planning,
and maneuvering for leadership of the Soviet Union, the state and
Communist party approved construction of the Dniepr project in November 1926.[253]

As in 1920, electrification led economic planning; this time it had to
wait for the other sectors of the economy. Instead of GOELRO preparing a national plan for electrification-based industrialization, Gosplan now handed Glavelektro its marching orders. The change in stat-
ure from 1920 to 1926 was great: electrification was no longer the basis
of the plan but part of a larger plan.

From the perspective of 1914 and 1917, an observer would speak
glowingly about Soviet electrification in 1925–26. The government
had created a national plan for electrification based on regional stations and hydropower. A formidable institutional framework existed
and utilities had surpassed their prerevolutionary levels of output.

From GOELRO's perspective in 1920, impressions were less favorable. GOELRO as a plan existed more on paper than in actuality, and
a leading opponent of Stalin headed the major electrification organization, which exercised less control than desired. Regional stations
had exceeded their cost and time schedules, budgetary and other constraints had scaled back plans, and local and urban utilities had rebelled against Glavelektro. Foreign transfers of funding and technology
remained far below expectations. Even worse, electrification was increasingly subordinated to the demands of heavy industry.

Financing proved a major weakness; European capitalists did not
succumb to the lures of Siberian wood, Baku oil, and Ukrainian grain.
Consequently, electrification had to compete for limited Soviet resources and never received the funding the GOELRO planners assumed. Until 1925, Soviet writers described the necessity of foreign
investment. After 1925, an emerging defiant attitude asserted that
"energy is the base of one of the most important commanding heights
which must be in the hands of the government."[254]

---

[252] "Ekonomicheskoe stroitelstvo," *Izvestiia*, 2 July 1925, 4; "The 'Dnieprostroy' Project," *Russian Review*, 15 October 1925, 409; "Po soiuz respublik," *Izvestiia*, 23 July 1925,
4.
[253] Rassweiler, *Generation of Power*, 30–58.
[254] Elektrobank, *Nekotorye itogi deiatelnosti*, 5.

Nonetheless, significant progress had been made. Soviet electrification rested on a firm institutional foundation, industrial users had increased, and new construction was incorporating the lessons learned from the first round of regional stations.[255] The official opening of the Volkhov station in December 1926 symbolized the first harvest of hydropower. Electric light and power grew faster than the rest of the economy, including railroads, which did not regain prewar levels of productivity until 1926–27.[256] This higher growth showed that allotment of resources to electrification had produced significant economic benefits. From an urban viewpoint, the key was not that electrification grew quicker but that its growth would have been greater and more geographically diffused had resources not been so skewed toward regional stations.

Just as 1914 proved as important as 1917 in shaping Russian electrification, so 1925 was as important for Soviet electrification as 1928, the start of the five-year plans. The halfway point in the NEP, 1925 marked a return to normalcy for utilities as interest shifted from recovery to expansion. Yet 1925 was also a year of change. Glavelektro's political and economic underpinnings, fluid through the early 1920s, were about to change again. Slow progress in constructing regional stations, the revolt of the utilities, and the increasing political significance of the peasantry forced reconsideration of a more locally oriented approach. The lack of political support from below, as demonstrated by the 1924 conference on electric supply, added momentum to revamping the GOELRO plan.

The basic problem facing the Soviet government and Communist party was that, if not controlled, most investment would go to the utilities of the second tier. Yet for large-scale industrialization, the country needed regional stations, especially in Moscow and Leningrad, still the country's major industrial centers. To strengthen its political base in rural Russia, the state wanted small rural stations, which were more costly to install and operate than central stations. To industrialize or socially transform, the government and party would have to violate the NEP's principles of decentralization, ideologically not a difficult task. The question was where to direct limited resources. Despite heated opposition, the state and party concen-

---

[255] A. Gorev, "Planovaia elektrifikatsiia," *Pravda*, 4 January 1925, 1; A. A. Gorev, "O sostoianii rabot po planovoi elektrifikatsii SSSR," *Elektrichestvo*, 1924, no. 12: 577–88.

[256] Roger A. Clarke and Dubravko J. I. Matko, *Soviet Economic Facts, 1917–81* (New York: St. Martin's, 1983), 10, 83, 84, 86; *Transport i Sviaz SSSR: Statisticheskii sbornik* (Moscow: Statistika, 1972), 91.

trated resources on regional stations. The rebellion of the lower tiers failed to change Soviet electrification policy. The urban utilities did not create a supportive political network throughout the government and party but operated within the existing administrative framework. By not creating a larger coalition, the utilities' decentralizing drive had to compete against the centralizing tendencies of Gosplan, Glavelektro, and Elektrobank without mitigating political and administrative allies. Worse, the Communist party's decision to force the pace of heavy industrialization increased the importance of regional stations at the expense of municipal utilities. Closure of the debate between conventional and regional stations ultimately—and effectively—came from above.

Glavelektro's 1926 revision shifted the original GOELRO goals while reaffirming its emphasis on regional stations. Most significant, electrification and its wide-reaching plans were now subordinated to less developed, more general plans for industrialization as part of the party's shift to centralized heavy industrialization and planning. Politically, electrification now was not the means to industrialization but a supporting foundation. Although it was discussed with GOELRO in 1920, only now was electrification's subordination fully accomplished.

The leaders of GOELRO, particularly Krzhizhanovskii, continued their work, but from the higher administrative heights of Gosplan in 1925. In a demonstration of the integration —and subordination—of the electrification enthusiasts into the government, their larger constituencies, responsibilities, institutional interests, and work differed from those of GOELRO in 1920. A lack of political support now distinguished electrification from technologies with greater support, such as its old nemesis, the railroads, and metallurgy. In the first years of the NEP, priority went to rebuilding the old, not boldly building the new. Lenin's weakening power until his death in 1924 deprived electrification of its most powerful patron. The association of Glavelektro with Trotsky weakened political support for the main agency for electrification, as did the continual changes in chairmen. Moreover, wide-ranging debates about the future economy and a struggle for leadership had filled the political vacuum of 1920.

The initial failure of the new state technology to fulfill its proponents' utopian visions had several causes. Very important was the neglect of electrification's dependence on other sectors of the economy, including the railroads, to provide utilities with resources to build and operate power stations. GOELRO based its plans on assumptions of centralized authority which bore little resemblance to the often strained industrial relations of the early 1920s. Its implementers had to contend with the NEP's more politically and economically decentralized environment. "Objective" economic factors, primarily a lack of expected domestic and foreign financing and foreign technology, further hindered growth.

The strength of the center consisted in directing resources into regional stations at a time when economic and political factors strongly favored investment in local utilities. The center exerted control not just by restricting local utilities but by eliminating potential industrial competition. From Moscow's point of view, denying factories the right to build or expand their stations to preserve a market for regional stations maximized limited resources. A planned economy in electrification meant not just favoring one type of power generation, but also stunting the development of alternatives.

The creation of organizations to plan, implement, and operate an electrified country demanded time and the resolution of issues ranging from profits to center–local relations. Some issues that appeared narrowly technical, such as standardization and inspection, actually involved conflicts between different political and economic interests over power and authority. Although local authorities won many issues, the overall control of the central organs, particularly Glavelektro and Gosplan, increased. Where Glavelektro faltered was in similar struggles against other central organs and interests.

The regional stations proved to be too much too soon. They consumed large quantities of money, materials, and skilled personnel at a time when all three were in short supply. Local stations fought regional stations on economic grounds for economic reasons: regional stations consumed the lion's share of limited resources. If the pie could not feed everyone, the utilities would fight for their fair slice.

Electrification had progressed greatly since 1913, 1917, and 1920. Electrical engineers could and did take pride in their accomplishments, even if they did not reach original expectations. The challenges facing electrification as the state technology and means of socialist transformation, however, threatened to diminish its importance. In 1920, electrification became *the* tool of the state. In 1926, it became only another tool.

CHAPTER 8

Conclusion: Shifting Grounds, Shifting Goals

THE TEMPO OF electrification in the Soviet Union increased sharply after 1926 as part of the state's renewed industrialization drive and the five-year plans. Electrification's share of the state budget grew from 68 million rubles (1.7 percent) in 1925–26 to 179 million rubles (2.2 percent) in 1928–29. During this period, state funding for industrialization quadrupled from 220 million rubles (5.4 percent) to 973 million rubles (11.8 percent).[1] Regional capacity and output grew sharply. By 1928–29, the Volkhov hydrostation generated 358 MkWh, 55 percent of Leningrad's 653 MkWh.[2] The first five-year plan, for 1928–32, reached GOELRO's goal of 1,750 MW from regional stations.[3] Fulfillment came from building much larger stations in a few industrial areas, not from building all the twenty-seven planned first-priority stations. The Kashira station, for instance, expanded from 12 MW in 1928 to 186 MW in 1932. The Red October, Shterov, and Shatura stations all expanded over the 100-MW level, as did the former 1886 Company Moscow station. The increasing concentration of generation in regional stations came at the further expense of municipal stations and the countryside. This realization of the GOELRO plan reinforced the centralized nature of industrial development and control.

---

[1] R. W. Davies, *The Development of the Soviet Budgetary System* (Cambridge: Cambridge University Press, 1958), 83.

[2] "Biulleten," *Elektrichestvo*, 1930, no. 3: 159.

[3] *Report of the State Planning Commission for the CPC of the USSR Summarizing the Fulfillment of the First Five-Year Plan for the Development of the National Economy of the USSR* (Moscow: Gosplan, 1933), 86–87; *Narodnoe Khoziaistvo SSSR za 70 Let* (Moscow: Finansy i Statistika, 1987), 32.

Equally important, regional stations spread from Moscow and Leningrad to the new industrial centers being created throughout the Soviet Union, including the Urals and Ukraine. These industrial centers and their supporting regional stations grew from the theoretical and technical foundations laid by GOELRO, Glavelektro, and Gosplan. Yet this major growth of electrification came at a cost to the electrical engineers and Soviet society. Glavelektro, now Glavenergo, moved from the VSNKh to become part of the Commissariat of Heavy Industrialization, completing the administrative subordination of electrification to industrialization. In a corresponding move, Prombank absorbed Elektrobank. The show trials and purges of the late 1920s and 1930s swept many engineers away, including Osadchii and Ramzin, as the party increased its control over all sectors of society. Radical social and political changes accompanied superindustrialization as the country encountered not only the five-year plans but revolution from above and below.[4] The vision of an industrialized economy powered by electrification had become real, but in a shape and through a process significantly different from what its promoters envisioned.

Even before the renewed emphasis on industrialization after 1926, Soviet electrification had advanced greatly compared with prerevolutionary utilities, although the gap with the West continued and even grew (see Table 8.1). Compared with 1913, the Soviet Union in 1926 had significantly advanced in the critical indicators of output and per capita consumption at the national and city levels. The gaps between the first-tier cities and other urban areas and between urban and rural Russia, however, grew.

Electrification was central to early Soviet industrialization, but not to tsarist calculations. Tsarist authorities viewed electrification as a normal technology unworthy of special attention, whereas the Communist party embraced it as the means to transform society socially, politically, and economically. But this sharp contrast should not obscure the deep tsarist roots of Soviet policy and personnel. Soviet electrification both continued and departed from tsarist electrification. Continuity existed in the problems utilities faced and in the leadership and goals of the electrical engineers. The major change was the

---

[4] See Kendall E. Bailes, *Technology and Society under Lenin and Stalin: Origins of the Soviet Technical Intelligentsia, 1917–1941* (Princeton: Princeton University Press, 1978); Sheila Fitzgerald, *The Russian Revolution, 1917–1932* (New York: Oxford University Press, 1982); Hiroaki Kuromiya, *Stalin's Industrial Revolution: Politics and Workers, 1928–1932* (Cambridge: Cambridge University Press, 1988).

Table 8.1. International electric generation and usage, 1912–26

Unit	kWh/person	Installed MW	Output (MkWh)
Countryside (1926)	.1	17	10
Russia (1913)	16	300	690
USSR	25	464	1,130
Kharkov (1914)	30	8	8
Kharkov	68	11	25
Moscow (1914)	92	78	182
Moscow	252	148	403
Germany (1913)	320	2,100	8,000
U.S. (1912)	500	5,221	11,569
Germany	600	5,938	11,521
U.S.	800	22,000	65,801

Sources: Countryside in 1926: G. Slobodkin, "Desiat let sovetskoi vlasti i elektrifikat-siia derevni," *Elektrifikatsiia*, 1927, no. 11: 15. Russia, Germany, and U.S. in 1912–13: L. Dreier, *Zadachi i razvitie elektrotekhniki* (Moscow, 1919), 8; "Statistics on the Operations of the Electric Light and Power Industry," *Electrical World*, 7 January 1928, 32; and B. R. Mitchell, *European Historical Statistics, 1750–1975*, 2d ed. (New York, 1976), 500. Kharkov and Moscow in 1914: "Statisticheskie svedeniia . . . za 1914 god," *Elektrichestvo*, 1917, nos. 4–6: 58–59, 75, 77. Kharkov and Moscow in 1925–26: S. A. Kukel, "Elektrosnabzhenie i elektrostroitelstvo v SSSR za 10 let sovetskoi vlasti," *Elektrichestvo*, 1927, no. 11: 366, and "Biulleten," *Elektrichestvo*, 1927, no. 4: 148. Countries in 1925: K. Gvozdev, "Dostizheniia sovetskoi vlasti v elektrokhoziaistve SSSR za desiat let," *Elektrifikatsiia*, 1927, no. 11: 17.

greatly expanded level of government interest and support for the new state technology, rooted in the increased visibility of electrification's economic importance and the technological utopianism of the new elites. A concomitant shift was the much greater participation of electrical engineers in the government and close ties between the electrotechnical community and the Communist party.

The war and revolutions dramatically changed both electrification and the environment in which it grew. As in the West, the war promoted centralizing tendencies and technocratic thinking in Russia. The subsequent Russian revolutions furthered these trends but also broke with the old regime. Only in revolutionary Russia, when the old tsarist power structure and technostructure had been discredited, did the goals of large-scale electrification and political power successfully converge.

Even if the tsarist or provisional governments had continued, the shock of World War I would have guaranteed postwar electrification higher political and economic status. Whether electrification would have become a state technology is far less certain. Redefining techno-logical choices implies redefining the political world. Revolutionary times often demand revolutionary technologies; conversely, the ac-ceptance of revolutionary technologies often demands revolutionary times. The regimes that overthrew the tsar embodied a revolutionary technological utopianism that meshed well with the enthusiasm of the electrical engineers. Electrification contained the promises of tech-nocratic thinking, planning, and societal transformation that consti-tuted the mindset of Communists and non-Communists in revolu-tionary Russia. In the uncertain environment of 1917–20, electrical engineers emerged as the only group with a firm plan for the future that went beyond rhetoric yet encompassed the goals of the govern-ment.

The Soviet government took a much more active role in planning, financing, and directing the course of the new state technology than its tsarist predecessor. The state took this step with the consent and urging of electrical engineers, who saw in the Soviet government a tool for accomplishing their goals. The Communist party and the state, in their turn, saw the plans of the electrical engineers as a means to achieve their vision of a socialist society. Despite this greater state attention, which provided electrification with significantly more resources and bureaucratic authority than in the prewar era, the new state technology failed to reach its early goals. Implementation suf-fered from problems of organization, the long construction times in-herent with regional stations, and events beyond GOELRO's control. Although written on a grand scale, the GOELRO plan did not allow for macrolevel disruptions like the continuing foreign hostility to the Soviet government or the introduction of the NEP. Nor, on the micro-level, could Glavelektro prepare for the death of electrification's most powerful patron, Lenin, or efficiently function with so many changes of its leadership. Equally important, however, was Soviet Russia's economic situation. Despite the claims about its potential to trans-form, electrification depended on the existing economy for fuel, mate-rials, money, and personnel. Before the war, the international electro-technical community supplied the equipment, capital, and personnel that developing Soviet Russia could not.

Two important issues for the history of any technology are the ap-propriate units of analysis and identification of the decisionmakers. In NEP Russia, was the correct timeframe the short-term approach of second-tier utilities or the long-term approach of regional stations? Were the criteria for investment priorities purely economic or also so-cial and political? Soviet Russia had a choice of three technological approaches toward electrification: conservative urban utilities, more adventurous regional stations, or radical rapid rural electrification. Each carried a different set of political implications. The issue was not of correct or incorrect choices but of the power to decide and implement. In the NEP's partially decentralized economy, the conservative path probably would have dominated as utilities expanded to meet demand and banks provided short- and medium-term financing. Building regional stations and long-distance transmission networks required substantial funding and a far greater reliance on foreign technology than the alternatives. Only the central government could have provided the necessary funding and waited several years before receiving a return on its investment. Although radical compared with prewar construction, this path was in the tradition of state-guided modernization. Radical rural electrification would have required a major reorientation of resources and priorities from large-scale industrialization and the urban polity to the countryside.

Supported by the electrical engineering leadership, the government chose the most politically advantageous path, which centralized decision making and concentrated resources in regional stations. The centralization of authority not only favored regional stations but also reduced the resources available to conventional and rural electrification. Thus, electrification as the state technology had adverse consequences for the second- and third-tier utilities. In a continuation and expansion of tsarist industrialization, economic and political elites chose a technological path that supported their power.

Judged by short-term economic criteria, both regional and rural stations were poor investments: the former demanded massive investment that would not return a yield for several years, and the latter were underutilized. Politically and socially, however, the rural and regional approaches appeared to be good, even necessary, investments. Regional stations seemed essential for the industrial transformation of the economy and rural stations seemed necessary for the social transformation of the peasant economy. In both, the state's priorities overrode conventional economic concerns, although its emphasis on regional stations reinforced the centralized bias of the new government and the "bigger is better" thinking of mainstream engineers. The Russian "bigger is better" school of utilities succeeded not by evident technological and economic superiority but by a political alliance with the state.

The concept of social and economic transformation raises two questions about deploying electrification for development. Was electrification the best investment to modernize and industrialize Russia? What was the best way to electrify? The political struggle over the future of industrialization answered the second question in favor of groups supporting centralized development. The issues of control and direction were discussed and debated. In contrast, advocates of electrification rarely asked in the open literature whether the resources demanded by electric light and power might be better invested in other services, such as medical care. Was electrification a poor investment and an improper technology for rural Russia? Were limited resources—financial, material, and human—best invested in revolutionary or evolutionary technologies? Should the first priority for modernizing the rural world be better plows, healthier horses, easier transportation to the city, or electric lights?

Rural electrification posed one of the great "what ifs" of the NEP years. What might have happened if the Communist party had attempted a "permanent revolution" by electrification in the 1920s? What would have happened if the state had dedicated to rural electrification a fraction of the resources spent on regional stations? What if the party had promoted a voluntary, cooperative-oriented collectivization, based on small power stations, in the mid-1920s instead of the violent collectivization of just a few years later? What would have happened if the 25,000 vanguard Communists sent to assist collectivization in 1929 had instead been quickly trained mechanics sent to electrify the countryside and give the peasantry a positive inducement to collectivize?[5] Similar efforts to "tractorize" agriculture in the late 1920s promoted the dual goals of collectivizing and increasing the productivity of agriculture.[6] Would electrification have established state–peasant ties and strengthened the hold of the Communist party in the rural areas? Would electrification have changed the prospects of the party in the countryside, or would a suspicious peasantry have rejected the electric lamp and motor of Ilich as an alien force, just like the "People's Will" half a century earlier?

The engineers also deserve scrutiny. The image of engineers and other professional groups as apolitical experts is only that. With the growth of technologies emerge groups of technical professionals who try to manipulate society within the context of larger political struggles. On the basis of their special expertise, Russian electrical engineers tried to monopolize technical issues in revolutionary Russia and

---

[5] Lynne Viola, *The Best Sons of the Fatherland: Workers in the Vanguard of Soviet Collectivization* (New York: Oxford University Press, 1987).

[6] E. H. Carr and R. W. Davies, *Foundations of a Planned Economy, 1926–1929*, vol. 1, pt. 1 (New York: Macmillan, 1969), 199–207.

gain political power, just as American physicists manipulated the federal scientific establishment after World War II.[7] Struggles between a specialty group and a larger public over issues usually develop with the specialists' claim that, because the issue in question is technical and therefore nonpolitical, the specialtists should make the decisions. Furthermore, claims by outside groups—often with competing agendas—are seen to be inherently political, biased, and therefore bad. Similar battles raged in the tsarist government as the ministries of finance, trade and industry, and internal affairs vied for dominance in economic development. A group of electrical engineers used the October revolution for their own revolution—and won, before losing to other economic and political forces. The series of trials, purges, and other coercive measures in the late 1920s and early 1930s eliminated any possibility of engineering groups holding power independent of the Communist party, even as the party supported technocratic projects.[8]

One key to the postrevolutionary establishment of far-reaching plans for electrification was the political success of the electrical engineering community in linking the two worlds of political power and electrotechnology. Gleb Krzhizhanovskii was not so much the hagiographically heroic individual as the right person with the right connections at the right time in the right place. Like Vannevar Bush, the organizer of the successful Office of Scientific Research and Development in World War II,[9] Krzhizhanovskii had strong links in the engineering and political spheres at a time when need—and opportunity—beckoned. Unlike Bush, Krzhizhanovskii lost his political benefactor soon, and the electrical engineers never succeeded in achieving a dominant position in the government. If the new government had not moved the capital from Petrograd to Moscow, could Krzhizhanovskii have created a GOELRO? Would the Petrograd leaders, such as Osadchii and Shatelen, have played larger roles? Or would GOELRO have been created at all? As Sapolsky's study of the Polaris program has shown, convergence between technological and political opportunity is not enough; success requires leaders "extraordinarily skillful in the art of bureaucratic politics."[10]

[7] Daniel J. Kevles, *The Physicists: The History of a Scientific Community in Modern America* (New York: Knopf, 1978), 349–92.

[8] Bailes, *Technology and Society*, 118–21, 418–20.

[9] Kevles, *Physicists*, 291–301, and Vannevar Bush, *Pieces of the Action* (New York: Morrow, 1970), 31–68.

[10] Harvey M. Sapolsky, *The Polaris System Development: Bureaucratic and Programmatic Success in Government* (Cambridge: Harvard University Press, 1972), 253–54.

Hughes's network approach has served to explain Russian electrification fairly well, but his approach can be extended. People do build systems, but systems that embody a large set of political as well as economic and technical factors. Hughes's cross-national comparisons of Germany, the United States, and the United Kingdom involved three countries with established economic, administrative, and technological infrastructures. Russian electrification suffered from infrastructural problems, political as well as financial and institutional, which the developed, industrialized countries of the West did not have. The failure of invention in Russia illustrates some of these problems, but the difficulties of technology transfer illuminate them even better. Technology transfer is not a passive act but a negotiated process in which the technology and environment are reconfigured to accommodate each other. Its success can rarely be taken as given and almost always involves recognizing the transfer's political and social dimensions.

Technologies tend to reinforce political and social patterns. The acceptance and diffusion of some technologies may indicate more about the society than about the technology. The slower diffusion of tsarist electrification and quicker Soviet embrace of centralized regional stations compared with the West have told us a great deal about the economic and political dynamics of Russian society.

In this book I have used comparisons to highlight similarities and differences between the West and Russia. Most comparisons are quantitative—per capita consumption, utility statistics—but qualitative comparisons remain vital to the identification and understanding of the history of technology. Statistics illuminate but do not elucidate. Numbers are constructs too; they cannot easily describe how random events, ranging from a good harvest to a chronic illness, shape the history of technologies. We tend to focus on a variety of specific factors rather than broader, less quantifiable issues. In our models and analyses of electrification, we must consider the larger environment, including institutional, financial, technological, and educational factors. Future research might compare Russia to India or to other less developed countries as well as to the West. The comparisons and contrasts should prove quite illuminating. Disentangling and understanding this weave is the joy of history.

Index

Academy of Sciences, 159; move to Moscow, 145; and war efforts, 100. *See also* KEPS

"Actor network," 10, 94–95, 137–38, 151–53, 164, 179, 185, 190, 194, 256

AEG, 34, 38, 53, 86, 245

Aleksandrovsk, 173, 186

Alekseev, N. M., Gen., 48

Alexander III, coronation of, 19

All-Russian Central Executive Committee, 126, 157, 222

All-Russian Conference of Communal Electric Station Managers, 224, 226

All-Russian Electrotechnical Congress, 14, 23–24, 34, 128; 8th, 178, 180–85, 187, 221, 248

All-Union Conference on Electricity Supply, 248–50, 255

Alternating current (AC), 42, 46, 70
  1-phase, 50
  3-phase: diffusion, 26, 34, 50–51, 53, 55–56, 58–59, 70–71; hydrostations, 83
  struggles with direct current ("battle of the systems"), 50–51, 62

American Institute of Electrical Engineers, 24, 124

Archangel, 104

Army, 16–22, 33, 56–57, 83, 199; Gerschenkronian substitute, 16; Special Commission to Supply the Red Army, 145; wartime hydropower, 112. *See also* GAU

ASEA, 241–42, 244–45

Association of Industry and Trade, 102, 115

Azneft, 208

Baku, 27, 46, 80; civil war in, 130; consumption, per capita, 73; electrification of, 58–

59, 70; and NEP, 195, 208, 216, 218, 223; and oil, 63, 79; output in, 110; in wartime, 108–9

Ballod, Karl, 166, 168, 190; influence on Russians, 139

Bayernwerk transmission grid, 189

Belgium: traction companies, 76–77; wartime diplomacy, 106

Belgium Company for Electric Lighting, 54

Belyi ugol. See Shatunovskii, Ia. M.

Benkendorf Company, 58

Bogolepov, Mikhail I., 115

Bogorod, 73; district *zemstvo*, 81; electric stations operating board of, 145

Boilers, 134–35, 149, 209, 213

Bolshoi Theater, 1, 174

Borovev, B. E., 183

Brown-Boveri, 38, 216

Bukhgeim, E. O.: 1914 proposal, 93–94; 1915 proposal, 116

Bureau of the Unified Technical Organizations, 109

Bush, Vannevar, 264

Cadet party, 143, 166

Callon, Michel, 2

Carlson, W. Bernard, 34

Carr, E. H., 253

Cement, 170

Central Bank for Agricultural Credit, 237

Central Communal Bank, 227, 229–31, 233, 249–50

Central Industrial Region, 78, 122, 241; electrification proposals, 116–17, 141–42; GOELRO analysis and plans, 160, 171–73; industrial use in, 70; local fuels, 109, 134;

266